Islands
Beyond the Horizon

The life of twenty of the world's
most remote places

ROGER LOVEGROVE

OXFORD
UNIVERSITY PRESS

OXFORD
UNIVERSITY PRESS

Great Clarendon Street, Oxford, OX2 6DP,
United Kingdom

Oxford University Press is a department of the University of Oxford.
It furthers the University's objective of excellence in research, scholarship,
and education by publishing worldwide. Oxford is a registered trade mark of
Oxford University Press in the UK and in certain other countries

First Edition published in 2012

Impression: 1

British Library Cataloguing in Publication Data

Data available

Library of Congress Cataloging in Publication Data

Data available

ISBN 978-0-19-960649-8

Printed in Great Britain by
Clays Ltd, St Ives plc

For Jim, Geoff, Iolo, and Mary who accompanied me to different islands and to the memory of my son Ross (1963–2011)

CONTENTS

ACKNOWLEDGEMENTS

I owe a great debt to all those people who have made my visits to the islands in this book so memorable. As friends and companions they have enriched my life to a huge degree and filled my mind with endless indelible memories. They come from many walks of life—lighthouse keepers, islanders, colleagues, travellers like me, fellow staff on numerous expedition ships and boatmen who have ferried me to innumerable remote places.

I also gratefully acknowledge those who have given of their time and patience in helping to revive my ageing memory and enable me to construct the twenty separate island accounts. In particular I thank Susan Barr (Jan Mayern), Katrina Johannessen (Mykines), Dan Vice, US Agriculture Dept, Suzanne Medina, University of Guam, Haldre Rogers, Isaac Chellman, and Kaitlin Mattos, University of Washington (all Guam), Olivia Renshaw and Natasha Williams (Ascension), Natalia Menelan and Armando Santos (Fernando de Noronha), Ben Buxton (Mingulay), Jim Flint (Tristan Da Cunha), Salvar Baldursson (Vigur), Palle uhd Jepsen (Halfmoon Island), Vikash Tatayah (Ile aux Aigrettes), Gabrielle Coulombe, Geoff Morgan, and Iolo Williams (all Tuamotu Archipelago), Robert Burton (South Georgia).

I owe a particular debt of thanks to Sandy Crosbie (lately head of the Geography Department at Edinburgh University) for his help in guiding me through the origins and characteristics of islands.

Various people have been very generous in allowing the use of their photographs for inclusion in the book. In this respect I thank Vikash Tatayah, Palle uhd Jepsen, Olivia Renshaw (and Ascension

Conservation dept), Ben Buxton, Susan Barr, Kris Zawadka, Jim Flint, Suzanne Medina, Armando Santos, Tim Olson. Their credits appear by the appropriate photos.

Finally I thank my editor at Oxford University Press, Latha Menon for her patience, encouragement, and advice at all stages of the book and Emma Marchant who has struggled manfully with my technical inadequacies while processing the illustrations and ensuring the progress of the book in its latter stages.

PREFACE

I still recall the shock in 1951 when my parents suddenly announced that for my sixteenth birthday they were paying for me to spend a week at the bird observatory on Skokholm Island off the Pembrokeshire coast of West Wales. They knew my interest in birds and other wildlife, but this visit was beyond any dreams I could ever imagine and to this day I have no idea how they found the money, relatively modest though it may have seemed; we were not a family that afforded holidays. From our home in north Cumberland (as it then was) to the far reaches of Wales seemed to me like a journey to the other side of the world. The train travel was everlasting but the week on the island was idyllic. I made lifelong friends and accumulated a swathe of unique memories which remain with me to this day, crystal clear in my mind. Two years later I managed to talk my way onto Lundy Island off the Devon coast, free for the summer in return for repainting the old lighthouse. This felt like the life for me; I was even offered a permanent job as a shepherd on the island but managed to decline it. But this was a second taste of island life and I was hooked. Islands were for me!

More than half a century later, on one of my many visits to St Kilda, far out beyond the Hebrides, I was sitting alone in bright sunshine on the soft turf on the slopes of Oiseval, enjoying a pair of Arctic skuas circling silently overhead and watching a fulmar sunk low on its nest on a nearby cleit. It was one of those precious moments of quiet solitude that life on a small island so often makes possible, when all seems well with the world, worries evaporate and the cocoon of the moment seems all that matters. I lay back on the grass, watched white clouds drift past and mulled over the incredible fortune and privilege I have had during my life, to have visited so many of the world's far-flung islands, in all five oceans. There is always a keen excitement about a forthcoming visit to a new island, although it is not simply the desire to visit the islands themselves that is the lure but, for me, understanding the ways in which the human communities exist in far-flung places; in some cases, of course, they have failed to stay the course and the island has been abandoned. It is a simple step from there to exploring how human presence over the years has interacted with the wildlife, frequently to its detriment. Pondering these thoughts, it slowly dawned on me that I should gather together these island experiences and record them, before too many of the memories slipped further into the grey mists of my mind. Thus it was that a peaceful hour on the slopes of St Kilda stimulated the idea of this book.

LIST OF FIGURES

LIST OF PLATES

1 Wrangel Island
2 Chinijo Archipelago
3 Jan Mayen Island
4 Mykines
5 Guam
6 San Blas
7 Ascension Island
8 Fernando de Noronha
9 St. Kilda
10 Pico in the Azores

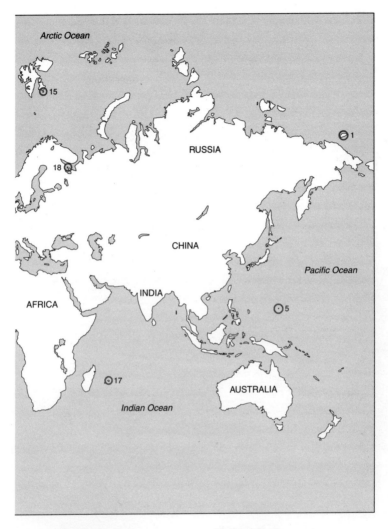

11 Tristan da Cunha
12 Vigur
13 Mingulay
14 South Georgia
15 Half Moon Island
16 Great Skellig
17 Ile Aux Aigrettes
18 Solovetski
19 St Peter and St Paul Rocks
20 Tuamotu Archipelago

INTRODUCTION

The book is a fairly personal account of twenty of the world's small, remote oceanic islands, all of which I have had the extreme good fortune to visit. Sometimes it has been to spend time there (Great Skellig, Fernando de Noronha), some to return to repeatedly (St Kilda, Halfmoon) and others to visit relatively briefly with the hope of one day going back (Pico, Wrangel). The question inevitably arises as to what defines remoteness in the context of islands such as these—or indeed others? For many islands it is self-evident, as in the Pacific where immense distances can separate an island from the nearest continent. However, we should remember that remoteness is not the same as isolation. Isolation is a matter of degree and can be found on an island in an inland lake or indeed in the middle of an urban surround.

Remoteness primarily involves distance but often also inaccessibility, not infrequently determined by severe weather or seasonal restrictions. Ice-locked Wrangel, storm-bound South Georgia, or even wind-swept, wave-lashed Mykines or St Kilda are good examples of that fact. Nowadays, with the sophistication of satellite communication, some of the islands I first visited decades ago, which were unquestionably remote and isolated, do not have the same feeling of isolation today. The family living on Toau in the Tuamotu Archipelago had no means of communication whatsoever when I was first there. Their island was remote and they were truly isolated; I trust that the same does not apply there now. With modern satellite communications, nowhere needs to be out of touch today although the physical problems of reaching distant islands often remain as challenging as ever. Readers may query the inclusion of two British islands within the definition of the book's title. However, I think if one asks the innumerable naturalists and travellers who have made repeated unsuccessful attempts to land on St Kilda, even in the summer season, they would concur with its inaccessibility! Similarly I include the magical island of Mingulay at the southern end of the Hebridean chain on the basis that it too has a strong story, a telling human history and is indeed far beyond most people's reach and very rarely visited.

What is beyond dispute, however, is that for many people small islands, wherever they are located, present an irresistible attraction. Size is undoubtedly an important criterion in relation to this 'island fever'. All the islands in this book are small—some very small— with the possible exception of Wrangel and Guam both of which might be described as medium-sized. Small islands are character- ized by their physical limitations, emphasizing their individuality and indeed defining their vulnerability. They reveal their identity and character more readily than large islands which cannot share the intimacy of the competing elements of land and sea. Great Britain is a large island, although the majority of people therein may not normally relate to the fact. In this respect it has no similarity to the feel of the small islands that lie in its surrounding seas. It is small islands like these, worldwide, that have such magnetic appeal.

There are no simple explanations for this island seduction. Nor is it easy to identify the factors which contribute to the allure because there are as many different answers as the number of people who are asked the question. Certainly on smaller islands, whether occu- pied by human communities or not, there is a feeling of being in a privileged world of one's own, a place of excitement and explora- tion, a private world with very visible boundaries. Each small island possesses its own character and wildness and is physically removed from the clamour of a busy outside world, bringing one tangibly closer to nature and possibly satisfying a deep, unrecognized primal need. Despite its remoteness, an island can paradoxically offer a feeling of both security as well as isolation; its boundaries are narrow and firmly defined. Added to these elements, many islands are visually attractive: in the tropics their captivating beauty can be as startling as a polished diamond on black velvet, but in higher latitudes, surrounded by wild waters, cliff-bound islands appear formidably awesome as well as starkly beautiful. For me there is one other important factor which contributes to the lure of islands. On all but one of those islands that I have been onto which have small human communities, there is a complete togetherness, a symbiosis of effort. Everyone helps everyone else and there is an atmosphere of implicit trust. I think of islands such as Mykines or Tristan da Cunha or perhaps, most notably, St Kilda where, within that recently extinct community, mutual reliance was the daily imperative of life and throughout its history, over the centuries, crime was an unrecognized phenomenon.

Whatever the nature of islands—distant, offshore, inhabited, uninhabited, tropical, or polar—their mystique and magnetism is ineffable, but the key to appreciating their individuality and understanding their distribution and physical features, lies in unlocking the history of their origins. Very large islands, such as Madagascar, are detached fragments of continents, separated many millions of years ago when Gondwanaland broke up and all the constituent continents except Antarctica, began to drift north. The British Isles likewise are continental in their geological origins. Worldwide, however, the vast majority of islands are volcanic in origin, including most of those covered in this book. The existence and distribution of volcanic islands is directly attributable either to the movement of the earth's tectonic plates or to 'hot spots'—relatively fixed plumes of hot rock in the earth's mantle. The former result in the creation of island-arcs as exemplified in Indonesia and in the Marianas or the nearby Carolines in Micronesia, while the latter produce islands or small groups that are scattered worldwide with Atlantic examples such as Jan Mayen, the Canaries, Tristan da Cunha, and Ascension.

Interestingly, single islands are relatively rare. Islands such as Jan Mayen and Ascension are exceptions. Archipelagos are the norm, either in linear groupings (e.g. Hawaii and Marquesas), as described, if their formation is the result of rapidly moving tectonic plates, or in clustered groups if the plate is motionless (e.g. Canaries, Azores).

Oceanic volcanoes, past, present, and future, are part of a timeless geological process wherein such islands are created and, in time, erode and eventually submerge again below the waves. The classic example of this process lies in the existence of innumerable atolls in the Indian and Pacific Oceans, where coral reefs have developed round the flanks of a submarine volcano as it emerged above the sea surface. The reefs continue to grow as the soft volcanic rocks of the cone are slowly eroded by the sea and eventually disappear below the surface, leaving the coral atolls as we currently see them, classically with a central lagoon encircled by the reef, defining the footprint of the extinct volcano. At the same time new islands are constantly forming over hot spots. I was fortunate enough to witness the violent emergence of Surtsey in the North Atlantic, off the south coast of Iceland in the 1960s. Surtsey provides a perfect demonstration of how islands come and go; sea erosion is already reducing it to nearly half its earlier maximum. Elsewhere other new islands continue to appear, as at Fukoto Kuokanaba (near Iwo Jima), also in the Solomons and in the Vava'u group in Tonga, not forgetting the on-going rebirth of Krakatoa in Indonesia.

The origins of some other islands are the result of the rise in sea levels following melting of the ice and the retreat of the last Ice Age 15,000 years or so ago. Initially levels rose rapidly but slowed down about 7,000 years later and have been relatively constant for the past 5,000 years. This slow rise of sea levels as the ice receded led to the inundation of the edges of continental land masses and created off-shore islands of different types, dependent on the nature of the pre-existing coastline. The Norwegian skerries and the chain of Lofoten Islands originated in this way, as did the Aleutian Islands when the Bering bridge linking Asia and North America was flooded. The Mediterranean Sea and its islands are a unique example of the same slow, long-term process. On several occasions in the geological past the Straits of Gibraltar have closed as the African plate pushed against the European plate. Some six million years ago this was the case and the Mediterranean, with its warm summer climate, completely evaporated. Eventually, however, around five million years ago, the Atlantic again penetrated the Straits and gradually flooded the basin, recreating the Mediterranean Sea and producing a rich scattering of islands, most notably the large archipelago in the Aegean Sea.

The size of an island is a major consideration in determining the scope, variety, and populations of the wildlife that exists there. Large islands develop their own local climates and range of habitats, coastal, inland, humid, dry, or altitudinal. Area is therefore a key factor in determining the number of species that will occur naturally on an island. On Madagascar, which has none of the large primate species present on the nearby African mainland, the lemurs effectively replace them and have evolved into some 100 species and subspecies through adaptive radiation. Small islands, such as all those covered in this book, can boast no such scale of variety.

These facts accord with the basic theory of island biogeography which recognizes that on larger islands there is a wider range of species while, conversely, smaller islands have less diversity. A compelling question is how remote islands first became colonized by species of plants and animals. The emergence of Surtsey as a result of a continuous eruption between 1963 and 1967, has given scientists a perfect opportunity to monitor the arrival and expansion of pioneering species on a new island 30 miles off the coast of the Icelandic mainland and 14 miles from the Westermann Islands. It provides a living laboratory to demonstrate how successful and effective primary colonization can be. By the summer of 1964 several species of flies and butterflies were already found there. In June 1965 while the island was still erupting, the first vascular plants appeared and seals were using the shoreline and have subsequently bred there. Ten years after the first eruption the vascular plant community numbered thirteen species and much of the island was clothed in mosses. At least ninety-one species of birds have been recorded on Surtsey but the first species to breed were black guillemots and fulmars in 1970. By 1986 a gull colony had established and soon expanded to comprise five species. Eighteen years later the colony had increased to well over 300 pairs, with clear evidence that their presence was already benefiting the vegetation. The fact that gulls produce many droppings, import food items, and regurgitate remains, as well as bringing ashore nesting materials, has produced soils with high nutrient values.

These factors have also been important in the development of plant and invertebrate communities in that part of the island. Regular monitoring of the island continues to record increasing species of plants and insects as the biodiversity of the island continues to develop. Bird species too have colonized. Before the end of the first decade of the twenty-first century, greylag geese, golden plovers, ravens, meadow pipits, and white wagtails were breeding successfully, together with at least ten pairs of snow buntings. Further species of terrestrial birds are anticipated to add to the breeding list in forthcoming years.

It is fair to recognize that the example of Surtsey relates to an island that is only a modest distance from a mainland and that colonization is therefore likely to be speedier there than on emerging islands that are more distant from land. However, the principles are the same and, in geological terms, time is on their side. Plant seeds and insects are wind blown across the oceans or sea-drifted on flotsam, by which means amphibians and terrestrial reptiles have also proved fairly successful accidental pioneers. Birds on the other hand—together with sea mammals—have the ability to arrive under their own steam and birds at least are recognized as being efficient accidental distributers of seeds and other minute organisms. Once a vagrant pioneer species has successfully established on an island, especially if it is isolated from the probability of further arrivals of others of its kind, it is likely to evolve and speciate, eventually producing a distinct subspecies, identifiably distinct from the original parent stock, and finally, with time, becoming a separate new species. Islands around the world thus exhibit endless examples of this and demonstrate gradual evolution into endemism. Almost all the accounts in this book mention endemism as a recurrent theme and the conservation of such species has become one of the main drivers of modern wildlife conservation initiatives on many remote islands. Flightless endemic birds have characterized many isolated islands, the dodo of Mauritius being the most famous, although a large number of islands have boasted their own endemic species of rail, most of them having become flightless as they are ground-dwelling and face no natural predators on these remote islands. Various other examples exist and are mentioned in the text, but it is not of course only birds to which endemism applies. The Canary Islands support a glorious range of some 500 endemic plants and the Galapagos (not covered in this book) are world renowned in this respect for the range of their indigenous species. There are other examples involving distinctive mammal species on island groups—usually bats, as they have the means of arriving on oceanic islands without outside assistance.

The only factor, other than natural disasters, disrupting the balance of wildlife communities on islands has been the arrival of man. The conventional wisdom has long been that Polynesian colonization of distant Pacific islands dated back to at least 1000 BC, with the most remote ones such as Easter Island reached by 400 AD, 300 years before Tahiti was discovered. However, recent research (Wilmshurst *et al.* 2011) has shown that colonization was in fact much later than this. They have shown that the eastwards colonization from Tonga/Samoa areas, 1,200 miles to the Society Islands and Tuamotus, actually took place some time between 1025 and 1121. The most remote groups such as Hawaii and Easter Island are now thought to have been discovered in a fairly brief pulse of human migration well over a hundred years later than this. Whatever the timeframe, these migrants were undoubtedly responsible for much deforestation and the extinction of native vegetation. Furthermore the first Polynesian rats to find their way onto such islands clearly arrived with the colonists and began the reduction and gradual elimination of ground-living species. It is apparent that ecological damage in these areas has taken place in a considerably shorter time than previously thought.

However, in considering the reasons for the declining fortunes of long-established wildlife communities on islands throughout the world—inhabited or uninhabited—no factor has been of greater consequence than the arrival of European man. His presence has been felt island by island in a myriad of destructive ways. Native vegetation has been eliminated and either replaced by introduced crops (e.g. forest clearance on Mauritius replaced by sugar cane) or the ground has simply been abandoned to unsustainable grazing (Tristan da Cunha and Ascension). Many species of birds have been exploited for food (the great auk was exterminated partly for that reason) and although there are a few examples of long-term sustainable harvesting of island birdlife, such as on St Kilda and Mykines, in most cases man's arrival pushed many species down the road towards extinction. Turtle harvesting worldwide has had this effect and continues to be a major problem in many parts of the tropics to this day. On islands such as South Georgia and Jan Mayen, enormous numbers of sea mammals were slaughtered to satisfy the demands of European and American markets. Nonetheless, the most widespread catastrophic factor has been the accidental importation of rats—Polynesian rat, black rat, and brown rat—onto islands across the world. The early European sailing ships were plagued with rats and it was an inevitability that they would escape from the ships at some stage and relentlessly colonize the new lands. Because of their distance from the nearest mainland, oceanic islands had no natural land predators and therefore their native wildlife had no defence against the arrival of rats or other introduced mammals. On innumerable islands the answer to growing populations of rats was to introduce cats or dogs in an attempt to control them, but in fact this achieved nothing other than making a dreadful situation even worse. Pigs and goats were brought onto many islands to provide ready meat and over the years they have destroyed much of the natural vegetation.

The period between the European exploration of the oceans in the sixteenth century (the age of discovery) and the latter half of the twentieth century was a long one of careless exploitation of plants, birds, and other animals on innumerable islands around the world. The toll on their wildlife has been incalculable, as is clear from many of the individual accounts in the following chapters. However, in the early years of the twenty-first century, there is considerable encouragement to be found in the increasing number of initiatives being undertaken to repair much of the damage, even on some of the most far flung islands. Apart from the initiatives mentioned in the individual accounts, there are many others to commend. Several islands around the UK have been successfully cleared of rats and the UK government is slowly taking an increasing role in restoring wildlife habitats and populations in its overseas territories. Other governments are undertaking similar steps. Some of the work is not only expensive but also daunting; the programme to eliminate rodents from the whole of South Georgia is one example. In the north Atlantic the Chinijo Archipelago has recently been designated as the largest marine protected area in Europe and the establishment of the 545,000 km^2 marine reserve around the Chagos Islands in the Indian Ocean has created the largest such area in the world. It has pristine waters which support over 200 species of corals, 1,000 species of fish, the largest diversity of seabirds in the Indian Ocean and important populations of sharks, dolphins, and turtles. The damage man has wrought on distant islands over the centuries has been appalling. The future for some of them may slowly begin to look brighter.

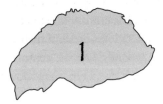

Wrangel Island

O f all the far-flung oceanic islands covered in this book, this is undoubtedly the most inaccessible. Nowhere else in the western hemisphere is there an island as impossibly icebound as to be unapproachable by sea in many years. It is not that it is hopelessly remote from a mainland, as many in the Pacific and Indian Oceans are, but rather that it is locked away in the fastness of the Arctic pack ice, visited by Russian scientists and a few intrepid travellers in the short weeks of thaw in mid summer. It supports a unique wealth of Arctic wildlife and has a discontinuous history of human occupation, traceable back to at least 1700 years BC. It also hides more recent events of unimaginable human trials and survival, locked away in the archive of its memories.

Wrangel lies in the frozen Chukchi Sea, a mountainous island 78 miles long, bisected by the 71st parallel and the 180° meridian, 100 miles north of the Siberian coast. As with most lands in the high Arctic, it supports an enormous range and biomass of wildlife. In many respects it is the Arctic parallel to the glorious richness on the much renowned island of South Georgia (Chapter 14) in the Southern Ocean, with its uncountable throngs of seabirds and legions of sea mammals. Remarkably, despite its extreme northern location,

the evolution of Wrangel's landform and vegetation was uninter-
rupted by the glaciation that razed and sculptured most of the Arctic
during the last Ice Age. As a result, its surface soils and topography
exhibit an exceptionally wide range of habitats which, in turn, support double the diversity of terrestrial species found in any other
polar lands at the same latitude. In fact the landscape today appears
much as it would have done in the Pleistocene, one million or more
years ago. Invertebrates of many taxa abound, including innumerable
spiders, bees, mites, moths, and beetles. Improbably, forty-two species of butterfly have been recorded, bearing testimony to the endur-
ability of organisms that we normally regard as fragile. Among the
wide range of resident butterfly species are fritillaries (*Boloria spp*),
blues (*Plebius aquila et al*), a copper (*Lycaena feildeni*), and a sulphur
(*Colias sp*). On a warm summer's day the butterflies bask in sunny
spots among the tundra flowers, with open wings soaking up the
fleeting warmth of the sun and gently raising their body temperature
for day-long activity. The diversity of tundra vegetation on the island
is remarkable for a location so far north. To date 417 species and subspecies of plants have been identified here, more than twice as many
as in any other polar area. It is principally on account of this unique
diversity that the island was designated as a World Heritage Site in
2004—the most northerly in the world.

However, Wrangel's uniqueness extends far beyond botanical
richness and invertebrate diversity. This is the site with the greatest
concentration of polar bears in the world. Some 500 females den on
the island each winter (a density of 15–30 bears per square mile (6–12
bears per km²) in some areas) and a further 100 females den on
nearby Herald Island. The world's largest population of Pacific
walrus—upwards of 100,000—live here and are the most numerous mammals to be found in the surrounding seas. They congregate
in huge rookeries on ice-free haul-outs, most notably the famous
site at Cape Blossom on the south-west headland and at Somnkeinaya
Bay, where the walrus are preyed on by the bears. Ringed seals,

Islands beyond the Horizon: The life of twenty
of the world's most remote places

which breed on the open ice, are the principal prey of polar bears, and bearded seals are numerous around the southern coasts, which are increasingly important feeding grounds in summer for grey whales migrating to the area from the coasts of California and Mexico. Arctic-dwelling belugas, the little white whales, and great bowheads are also found here with smaller numbers of humpbacks and fin whales which feed and migrate through the same waters. Arctic foxes are numerous and reindeer were first introduced by settlers in 1948 and their numbers have burgeoned into thousands. Twenty musk ox from Alaska were reintroduced in 1975 and their population too has increased to more than 800.

Wrangel was not firmly added to the world's known geography until 1867, hidden as it was in the vastness of the icy Arctic wastes. Nonetheless, there had been ancient rumours of the existence of a 'great land to the north'. Chukchi myth and legend tells of an erstwhile people migrating from the Siberian mainland to a land far to the north. Baron Wrangel, a Russian cavalry general of German origin was the somewhat improbable leader of an expedition surveying the unexplored Siberian coast and claimed a temperature-inverted vision of an island in this area in 1820. HMS Herald in 1849 and a German whaler in 1866 also claimed to have seen distant images of high land across the frozen seas. However, it was left to an American whaler captain, Thomas Long, eventually to confirm the island's existence and position in 1867, when he approached to within five leagues through the ice in the channel that now bears his name. The decades following its discovery were riddled with conflicting claims on the island by Canada, Britain, and Russia. The issue was finally settled in 1926 by the Russians not only raising their flag, but through the simple expedient of transporting fifty Chukchi people there to form a permanent colony at Ushakovskaya, thereby establishing sovereignty.

Even in an environment as unforgiving as this, man's presence in the twentieth century soon began to have a devastating effect on the wildlife, both directly and indirectly. Wrangel Island supports the only breeding population of lesser snow geese in Asia and the numbers here are prodigious and of outstanding international importance. Thousands of pairs breed colonially in the mountainous interior, in valleys of the Tundra River and Mammoth River, where densities reach as high as 5,000 pairs per square mile (2,000 pairs per km²). The settlers lost no time in harvesting birds and eggs to feed both people and their sled dogs and there was uncontrolled hunting in the years after their arrival. The introduction of reindeer also had an immediate effect as the grazing herds began to damage the fragile tundra ecosystem. Furthermore, as the herds moved around the island they trampled goose colonies and ate the eggs. Arctic fox numbers tripled on the back of human refuse and waste, thus putting further pressure on the geese and other ground nesting birds. By 1960 the number of geese had fallen catastrophically to no more than 50,000 pairs and a refuge area was established to stop the decline. The numbers recovered steadily thereafter and in 1976 the whole island was declared a State Nature Reserve, first and foremost on account of the lesser snow geese which continue to increase their numbers and are now well in excess of 250,000 pairs. The little settlement at Ushakovskaya increased in size after the Second World War to about 1,200 souls, with the addition of a military presence and a polar research station. However, by the end of the Soviet era and the withdrawal of the military, the numbers fell and the settlement was eventually abandoned. Now the island is only populated seasonally by a few research workers and reserve staff. Nonetheless this twentieth-century story of occupation is only a part of the human history of Wrangel, because it is now known that the island, although never occupied by Yup'ik (Siberian eskimo) peoples, was inhabited far back in prehistoric times. Archaeological evidence points to early human settlement and has discovered ancient stone and ivory tools, carbon dated to c.1700 BC. At that time, as well as the native wildlife that we recognize today, these early people were contemporary with the last of the woolly mammoths, many remains of which from the same period have been found on the island. This was evidently an isolated mammoth population, far removed from the mainland, but there was also a dwarf subspecies (*Mammuth primogenius wrangelensis*) here, some 25 per cent smaller than the mainland animals and unique to the island. The remains of both species have been found in abundance on Wrangel, together with ancient spears, etc., but it can be no more than conjecture as to the part early man played in the extinction of either animal on the island.

More recent human events on Wrangel illustrate the hazards of travel in the dangerous polar waters around the island. In 1879 the 428 ton USS Jeannette sailed through the Bering Straits on an expedition attempting to find a passage to the North Pole. By 4 September she had sighted Herald Island, 60 miles north-east of Wrangel, but was almost immediately trapped in pack ice. For the next 21 months she drifted in the pack before eventually breaking up, leaving the crew with no option but to trek an unknown distance across the ice to the Siberian coast where, even then, the chances of rescue were remote. Most of the ship's complement died at different stages and the journey of the remainder, across the frozen tundras, was an epic of survival which only thirteen men completed.

Another disaster acted out on Wrangel was that of the Karluk in 1913–14. The ship, an America-built brigantine under the Canadian flag, was on an Arctic expedition led with shattering incompetence by Vilhjalmur Steffansson. It was heading north to carry out scientific work in an area of the Beaufort Sea. However, this ship too was soon trapped fast in pack ice and began to drift north-west. Steffansson decided to leave the ship with five other men to go hunting, but as the Karluk drifted steadily north-westwards, he failed to find the ship again and was eventually forced to cross the ice to the mainland in Alaska. Meanwhile the Karluk, locked in the ice, continued north-west for the next four months before the ice drove a large hole in her hull and the crew had to abandon ship and make camp on the ice. Here, in the depths of the Arctic winter the Captain, Robert Bartlett, organized the building of igloos, in which the crew survived until the partial return of daylight in February. Bartlett then determined that they would set off to walk southwards towards Wrangel Island. Four men, who disagreed with his plan, set off on their own, and were never seen again. Two other men sent by the captain went ahead of the rest to look for Herald Island, which he thought was 50 miles or so to the south. They too were not seen alive again although their bodies were found on Herald Island fifteen years later, proving at least that they got there.

The story of the following eight months from the departure from Shipwreck Camp began with a treacherous 100 mile, 22-day walk across the crumpled ice hauling all their gear is one of desperate privation which they all suffered. Inevitably some died while others successfully fought to stay alive. Six days after eventually staggering to Wrangel, the incomparable Bartlett set out again with an Innuit companion to walk 200 miles across the frozen sea to Siberia, a remarkable journey in itself which he completed in less than three weeks. After refreshing and feeding properly in a Chukchi village, Bartlett immediately set off again with dog sleds for another 500 miles across unforgiving tundra, to the Bering Straits to seek a rescue ship. Arriving towards the end of April 1914, it was a month before a ship could be found, and it was not until 7 September the same year that the skeleton party of eleven surviving men was taken aboard, ironically by a ship that happened to be passing shortly before Bartlett arrived. The survivors had existed on whatever meagre fare Wrangel could provide in the latter days of winter and on into summer. They lived on roots, skins, and whatever occasional animals or birds their Innuit woman could catch. Those few who lived to be rescued were little more than skeletons and their story is yet another among the epics of polar survival. Wrangel is a seriously forbidding environment most of the year—there are, for example, only two to three weeks when it is frost-free—but, for the brief daylight weeks of summer, it grudgingly shows a softer side.

To stand on the coastal flats of Wrangel on a sunlit day in mid summer is to lift the veil to a naturalist's paradise. The tundra is a tapestry of subtle and varied tones of browns, a rich underlay for the carpet of myriad flowering plants and the buzz of insects. Most of the plants grow tight to the ground, finding what shelter they can from desiccating Arctic winds, for it is the savage wind of winter that is the killer, more than the plummeting temperature. Polar willows and birches, often ancient trees growing horizontally along the surface at no more than 1/400inch (1/10mm) annually, keep many of their leaves over winter to save time in the brief days of summer next year. Beautiful Arctic poppies in shades of yellow are everywhere— no fewer than seventeen species are known, five of them endemic to the island, including *Papaver gorodkovii* and *P. lapponicum*. The four petals of each flower form a tight parabola which carefully follows the sun throughout the twenty-four hours of daylight so that the sun's rays are carefully directed to the reproductive centre of each flower. Other plants grow in tight cushions which trap tiny pockets of warm air. The tufted heads of pale paintbrush push up through carpets of pink dryas and arctic saxifrages (*Saxifraga spp*). Like most other plants in these environments, they are perennials as there is little time to flower and set seed as annuals; often they reproduce by sending out runners in the shallow band of soil above the permafrost. There are pasque flowers whose soft hairy surfaces have again evolved to trap grudging warmth and whose flower heads close up at night to lock in any pollinating insects at temperatures 8–10° warmer than the outside. Alpine arnica, thrift, bladder campion, and arctic buttercups (*Ranunculus spp*) are among many others that have evolved individual strategies to deal with the harsh conditions. Marsh marigolds and cotton grass (*Eriophorum spp*) flourish in the wetter areas; the cotton grass unusually develops a set of leaves as autumn approaches which give a flying start to photosynthesis the following summer. The moss campion and dryas both grow in tight cushions; dead leaves are not dropped but accumulate to help increase the cushion effect and provide shelter for the new growth with a temperature inside the cushion 10–15° warmer than the outside air. Only in the slightly warmer interior of the island, among sheltered river valleys, do willows grow as much as a metre in height. The island's Pleistocene credentials are emphasized by the existence of three relic endemic plant species from that era, the entire world population of all three being found only in small areas in the upper reaches of the Neizvestnaya River valley.

Across the tundra, tough little snow buntings flit from one stony slope to another, a Lapland bunting sits atop a low willow bush and the phantom paleness of an Arctic redpoll is a surprise. It was here, as I stood one day soaking up the kaleidoscopic carpet of plants, that I heard one of the stirringly evocative calls on the high Arctic breeding tundras: a grey plover (the Americans more appropriately call it black-bellied plover) in magnificent silver and black summer plumage, on a rise in the ground 110 yards (100 metres) away. It called with a fluty, melancholy, bell-like note. It called again and then again. A while later, I was evidently walking too near a nest, for the female lured me away with a drooping injured-wing display. Grey plovers are not alone, for there are a dozen other wading species nesting on these tundras, the most numerous of which are turnstone, knot, pectoral sandpiper, and dunlin. As I watched the grey plovers they became suddenly agitated and took to the air in earnest to harry a passing long-tailed skua—the most graceful and streamlined of its family. Together with pairs of gyr falcons and peregrines, the skuas are one of a handful of avian predators that breed on Wrangel to take advantage of the seasonal bounty of available prey. In years of lemming abundance both skua and snowy owl prosper enormously. The owls may rear as many as nine young in a nest and in such years up to 400 pairs can breed on the island. In the lean years, however, when the usual high numbers of collared lemmings and Siberian lemmings collapse, the owls may not breed at all.

Among all this richness of animals and plants for which the island is famed, the snow geese are the emblematic species. They arrive in long skeins in May, as the snows begin to thaw, although if the timing is wrong or the thaw is late, they must wait and live on their fat reserves for as long as necessary. In July and August when the eggs hatch, many of the pairs lead their goslings for 50 miles or more, out of the high river valleys, where they breed, and across the tundra to the coastal plains and lagoons, where they are safer from predation by Arctic foxes and wolverine and where they prepare for the daunting migratory journey to the South.

As the sun weakens, summer slides into autumn with a warning suddenness and Wrangel begins to empty. The myriad birds move away: song birds, wading species, wildfowl, avian predators, and the throngs of seabirds from the breeding cliffs at the eastern and western ends of the island, all depart for the south. Vast flocks of

FIGURE 1 Musk ox.
© The author.

lesser snow geese fill the sky with their wild, haunting calls as they head south along the ancient corridors of time towards their winter grounds in the valleys of the Frazer and Skagit rivers and the lowlands of central California. The sea they leave behind freezes over and seals and walrus are forced to abandon the shores. As the days shorten and the snow returns, the female polar bears make their way to the predestined hollows on the hillsides where they will submerge themselves in the deepening snow for the winter and give birth to their young. Among the 62 species of birds which breed on Wrangel each year, only two, snowy owl and raven, actually stay put to survive through the bitter cold and darkness of the Arctic winter.

One lasting image remains as Wrangel closes down. As a bone-chilling, knife-edged wind coming down from the mountains bears the first stinging squalls of snow, a group of five tough, shaggy musk ox stand, short-legged and defiant, outlined against the fading skyline: winter sentinels guarding the frozen land, sealed in ice and snow awaiting the promise and rebirth of distant spring.

Chinijo Archipelago

The second half of the nineteenth century and the first two decades of the twentieth were exciting periods of extensive ornithological discovery by European naturalists travelling to places that had barely had any of their wildlife described previously. It was the heyday of Victorian exploration and many new species were added to the lists of known birds in different parts of offshore Europe and the Arctic as a result of these expeditions. Among these areas, one of the most popular destinations was the Canary Islands. The nine islands varied greatly from one to another and each had an equally rich and differing avifauna. Over seventy written accounts in books and scientific papers, covering that period, testify to the number of visits and the amount of work that was done on the islands, mainly by ornithologists from Germany and Britain. In the early years of the twentieth century a notable British ornithologist, David Bannerman, followed in the footsteps of others before him in spending time there and, over a span of several visits, he made important discoveries on all of the islands from La Palma in the west to Lanzarote and its satellite islets in the north-east.

In 1913 Bannerman was on a pioneering exploration of the remote eastern islands. On the morning of 3 June, on the little desert island of Graciosa where he was camped, he was alerted by a local fisherman whose brother had, the previous evening, seen an example of the rare and almost legendary Canarian black oystercatcher on the far side of the island. In great haste and a fluster of excitement, Bannerman and his guide hurried to the spot where the bird had been seen and, to their delight, they found it still there, feeding among the black laval rocks a hundred yards (90 metres) across the shore, distinguishable only by its brilliant red bill. Moving stealthily round the dunes to get a closer view, Bannerman rose to his knees, fired and proudly killed the bird with one shot. This was the last of its species ever to be seen by ornithologists. It is easy to be critical of such actions a hundred years later but it should be remembered that in the decades around the turn of the twentieth century, killing specimens such as this to produce museum specimens was regarded as normal practice in helping to define new species and complete the knowledge of the birds of different areas. It was a perfect example of the hackneyed adage, 'what's hit is history, what's missed is mystery'.

Graciosa is the only islet in the north-east that supports a small human population. Judging by photographs taken in 1913, little seemed to have changed over the sixty years following Bannerman's visit and up to the time of my own first experience of the island with colleagues in the early 1970s. The same group of white-painted, flat-roofed cottages comprised the little village of Caleta del Sabo on the eastern shore, with the fishermen's boats rocking gently on the swell or hauled up on the sun-lit beach. Outside the cottages the sandy streets were still regularly brushed clean of the previous day's foot-marks in the early morning and there were no wheeled vehicles of any description to disturb the silence of village. Pale-plumaged yellow-legged gulls patiently patrolled the harbour and the deep-throated calls of a pair of ravens came and went in the wind around Montaña del Mojon in the middle distance. The hinterland of the island was certainly the same arid wilderness of blown sand and dust as Bannerman had photographed, with a scatter of desert vegetation, exciting for those who enjoy desert environments and at the same time a poignant example of desertification caused by endless generations of overgrazing by flocks of free-ranging goats.

Any mention of the Canary Islands readily conjures images of sun-drenched holiday destinations, sandy beaches and ever-cheerful overflowing crowds of north European tourists. These are islands of perpetual warmth and pleasure seeking and the fact that their environments have been drastically altered through the development of holiday facilities and expanding agriculture, at a cost to their previous wildlife diversity, is beyond the scope of this account. However, in recent decades the tourism focus has even expanded from the two main holiday islands of Tenerife and Grand Canary to embrace the smaller islands to the west as well as the great desert islands of Fuerteventura and Lanzarote to the east. Graciosa, the nearest of a group of three small waterless islands, lies across the narrow Rio Strait beyond the far headland of Punta Fariones at the north end of Lanzarote. This very special group of remote islands exists in a world far removed from the crowds of sun-seekers. They are of little interest for tourists and, with one exception, are little visited. They are all true desert islands, a mere 100 miles (160 km) across the sea from the barren West African coast. They lie in the eye of the dust-laden north-east trades, blowing off the desiccated Saharan hinterland of southern Morocco and Western Sahara; islands of perpetual dry, choking wind. Even the recently invented name of the group, the Chinijo Archipelago, is unfamiliar to most people. However, there is no doubting the biological importance of these far-flung islets and rocks, for they harbour the most important populations of several bird species in the Canaries as well as fourteen endemic invertebrates and a rich representation of sea life. They are a little known European treasury.

All of the islets betray their origins, with volcanic cones or the remains of them, being the dominant feature on each one. Graciosa, 5 miles (8 km) long, has four prominent cones, the highest rising to over 820ft (250m). Montaña Clara, a mere stone's throw off the north point of Graciosa, is little more than the shattered remains of one large volcano, half of which has long since detached itself and sunk into the depths of the ocean. It is virtually devoid of vegetation, has no fresh water, only a pocket handkerchief of flat ground and its whole surface is a wilderness of twisted lava, the legacy of the outpouring from its fragmented cone. A little farther out, and the most northerly point of the Canarian archipelago, the roughly circular island of Allegranza, 3.8 square miles (10km²) in extent, is dominated by Montana de la Caldera rising to 941ft (287m) on its western end, surrounded on its other sides by dusty lava plains. On one of these plains are the shadows of former cultivation strips, for even this inhospitable island was once occupied, not only by lighthouse keepers, but also by one family who eked out what must have been a hard won existence there, catching fish from the sea, taking birds from the land, and growing small patches of wheat. Two impressive sea stacks, Roque del Oueste and Roque del Este complete the group.

Graciosa has maintained a greater variety of wildlife than the other islets, but even here the present spectrum of species and numbers is undoubtedly a pale shadow of what the island supported in the past. Now, in the early years of the twenty-first century, the result of overgrazing throughout the centuries can be seen clearly. Standing on the slopes of the highest hill, Aguja Grandes, and overlooking the rest of the island, the transformation of its original vegetation cover can only be imagined. True desert dominates the whole of the island, low, flat, and dusty around the four volcanic cones. Only the hardiest of plants here survive the ravages of the goats and the parched conditions that characterize the island. The scattered plant community comprises only succulents, some unfamiliar, with no common English name—*Traganum moquinii*, *Zygophyllum fontansi*, and several *Euphorbia* species—with other more familiar succulents such as seablite, glasswort, and sea orache together with clumps of ferociously spiny *Launaea arborescens*. These, and a handful of other plants, are the stubborn survivors that have withstood the effects of centuries of grazing: living reminders, perhaps, of the 520 endemic plants within the remarkable Canary flora of some 2,000 different species. However, the damage to natural habitats throughout the islands has been so severe that almost 600 species of these plants are now regarded as threatened or endangered.

The modest range of terrestrial birds that inhabits Graciosa has changed little in the last hundred years, even if numbers have dwindled, and its composition reflects the desert nature of the island. Stone curlews breed in small numbers, pairs of Kentish plovers enliven the sandy beaches and a few trumpeter finches and spectacled warblers are found in areas of remnant vegetation in some of the dried up gullies. The most numerous birds are pale, desert-coloured Berthelot's pipits that run from one small eminence to another, scuttling across the ground and frequently showing a marked reluctance to fly. This interesting, if undistinguished bird is the smallest member of the worldwide family of pipits and wagtails and is a Macaronesian endemic found only here on the Canaries and on Madeira, on all of which islands it is common. It was named after a French naturalist, Sabin Berthelot, who was resident on the Canaries in the 1840s; his distinctive pipit remains a universal signature bird on all the islands, not least on Graciosa and the other islands of the Chinijo group.

However these terrestrial birds are only part of the story in the Chinijo Archipelago, now designated a Natural Park. There are several seabird species which breed on the islands in important numbers and have been annually exploited by man. Shearwaters and petrels all nest underground or in boulder hollows and crevices and only visit the nesting areas during the hours of darkness. Most numerous here are the Cory's shearwaters, dark brown birds with white underparts which soar and glide in characteristic undulating flight above the surface of the sea. They are large birds, weighing up to 2.5lbs (1kg) and have for many years been much favoured as food throughout their Mediterranean and sub-tropical Atlantic range. They breed on all islands in the Chinijo Natural Park and young birds prior to fledging were, in the past, harvested by the Graciosan fishermen, both for their own consumption and for sale in the market in Arrecife on Lanzarote in the days when Arrecife was no more than a small fishing village.

The shearwaters, 'pardelas' in Spanish, were taken on Graciosa for as long as locals can recall, although the long-term effect this may have had on the breeding population there is unknown, although it is likely that they were far more numerous in previous centuries than they are now. The highest numbers breed on Allegranza (which supports the second largest colony in the world) where all parts of the island provide sites for breeding pairs, wherever there are suitable hollows, crevices, or caves. The talus slope of huge boulders at the base of the one lava-sand beach is a breeding fortress for the birds. Attempting to sleep on the beach, with or without a tent, during the breeding season is an exercise in futile ambition. It is not only the night-long, ear-splitting caterwauling of countless scores of individuals returning from the sea, but also the physical bombardment by squadrons of birds coming and going that makes sleep no more than a wistful dream. To make the nighttime experience more memorable, the beach is a moving carpet of house mice which obviously prosper on the detritus and leftovers from the colony. Up to the middle of the last century as many as 8,000–9,000 young shearwaters were taken on Allegranza each year by the Graciosan fishermen, but even that level of harvesting probably had little detrimental effect on overall numbers. The 'pardelas' are long lived birds and can tolerate considerable predation. On Graciosa there was some recreational shooting of the birds at sea in September after the young had fledged, but it was small-scale and of negligible consequence to the population but has now stopped in any case.

Other shearwater and petrel species also breed on the islands. Small numbers of little shearwaters (more recently renamed Macaronesian shearwater) are to be found on each of the three largest islands, together with similar numbers of the all-dark Bulwer's petrel. It gives away its presence deep among the boulder beaches with its puppy-like call, 'yap yap', which bequeaths it the local name 'perrito' (little dog). Strictly nocturnal, they disappear from the nesting beaches before dawn and are often difficult to find out at sea. In the past they were reputedly more numerous than nowadays and were heavily harvested. Three storm petrel species breed on the islands, again in very small numbers, the British storm petrel, the Madeiran storm petrel, and the delicately beautiful white-faced petrel. Even on the rugged Roque del Este there is evidence of the former two species breeding. We found the remains of birds and recent footprints in soft dust at crevice entrances. The rock rises almost vertically out of the ocean, its surface extremely loose and friable, making it difficult of access and very dangerous to work on. Its twin peaks are linked by an extremely narrow, unstable ridge with sheer drops to the sea on either side. Its shattered nature is partly due to the fact that, in ignorance of its wildlife importance, it was briefly used as a target for naval gunnery training!

No such problems of access are found on Allegranza, once a landing is made on its beach, for the ascent over the block boulders to the plateau is relatively easy. The volcanic cone at the west end dominates the island, rising to a height of 941ft (287m) with a caldera that is over half a mile (800 m) in width. The remainder of the island is a mixture of lava-covered slopes, dusty plains, and low stony hills. It has the feeling of abandonment, strongly enhanced by the shell of the small cottage that still exists and the knowledge that a family once struggled and toiled here. Like Graciosa and Montana Clara, Allegranza is extremely arid although, with modern freedom from goats, its desert vegetation is better developed, with thickets of large *euphorbias* particularly notable. Apart from the one sandy beach, the coastline is ringed by precipitous lava cliffs and caves, much used by the resident rock doves, and stretches of black, rocky reefs on the shoreline—ideal areas in the past for the black oystercatcher.

At the eastern end, on Punta Delgada, is the Allegranza lighthouse. In the 1970s we were fortunate in tracking down the last two keepers who manned the light for twenty years from 1949, before it became automatic, and they were able to give a clear account of the status of many of the birds on Allegranza in their time. So far as the black oystercatcher was concerned, they were very familiar with the fact that it had been found regularly on the island in the past, although they never saw it themselves. They heard stories from older men that it was last seen up to ten years or so before they were the occupants of the lighthouse. In the past it had been seen frequently on the reefs around the lighthouse throughout the year, sometimes in twos or threes, although it had become less and less regular as time went by until the last one the old keepers ever saw was about 1940. In recent years the bird has been confirmed as a separate species and not a subspecies of the closely related black oystercatcher in South-West Africa, as was previously thought. However, the reasons why it should have declined and become extinct—not helped by Bannerman's efforts—remain something of a mystery, albeit its population on the eastern islands was probably never large. Only four specimens of this iconic bird remain, three in the British Museum of Natural History and the one shot by Bannerman in Liverpool Museum.

Another notable bird which occurs on Allegranza (and also on the other islets of the Chinijo Reserve) is the elegant and rare Eleonora's falcon. Eleanor of Arborea was a Sicilian princess in the fourteenth century, who, known not only as a formidable warrior, was also famous for enacting early laws for the protection of nesting falcons. She is fondly remembered now by the elegant falcon which carries her name. It is a particularly interesting bird, related to the hobby and the sooty falcon, and has a breeding range restricted to island cliffs throughout the Mediterranean with—somewhat strangely—outlying colonies beyond the Straits of Gibraltar, off the coast of Morocco, and in the Chinijo Archipelago. In the early 1970s we were able to show for the first time that there is a strong breeding population on these islands, probably as many as fifty pairs, including a measurable colony on the remote sea stack of Roque del Este. Apart from the falcon's strictly delimited breeding range and its elegance, dash, and panache, two other facts distinguish it from other falcons. First it has a unique migration, the entire population leaving the Mediterranean, Morocco, and Canaries at the end of the breeding season, and undertaking a 6,000 mile (9,500 km) journey for the majority to spend the winter in Madagascar. Until recently it was believed that the population, from all parts of the Mediterranean, and by implication the Canaries, took a route eastwards, then passing south through the Red Sea and onward down the east African coast. However it has now been proved by satellite telemetry that, more understandably, the birds are trans-continental migrants, crossing the Sahara and the equatorial rainforests and reaching Madagascar in November via Kenya and Mozambique.

The second unusual aspect of the Eleonora's falcons is the timing of their breeding cycle. Birds arrive at the colonies from late April but egg laying does not begin until late July or early August when most other land birds have finished breeding (our stay on Allegranza, 26–31 July 1970, coincided with the peak of the laying period). This means that, with a 28-day incubation period, young are in the nest from the first week or so of September until they fledge around mid October. This unusual fledging date is timed to coincide precisely with the period when the maximum numbers of song birds are passing through on their southerly migration, thereby providing a ready source of prey for the falcons which would not be available for them earlier in the season. During the rest of the year the falcons feed extensively on large flying insects but now, with young to feed, they

FIGURE 2 Eleonora's falcon.
© Wild Wonders of Europe/Unterthiner/Nature Picture Library.

convert to an exclusive diet of small birds. Examination of the prey remains around nest sites gives ample evidence of the wide range of European song birds that are taken. On Graciosa, in the weeks before the migrating birds appear and the young falcons have hatched, there is a clear southward passage of adults in the mornings and, particularly noticeably, in the evenings as birds move back to their territories on the outer islands for the night. This daily movement is presumably because the numbers of potential prey species is so low on the arid islands where they breed, that they have to cross to Lanzarote and possibly even as far as Fuerteventura to hunt.

Lightly built, fast and agile they may be, but they are bold in their attacks on much larger species in the vicinity of their territories. Safe in the reliance of their speed and agility, they boldly attack ravens, ospreys, and even Egyptian vultures, not infrequently causing them to dive into refuges in caves or other shelter on the cliffs.

The great biological importance of the Chinijo Archipelago is well understood and has been recognized in a suite of recent designations. Its bird communities may be the most obvious elements but, as mentioned, it also boasts endemic invertebrates and rich marine life. As well as protection afforded as the Chinijo Natural Park, it is listed as a Special Protection Area under European legislation (the islands are always illogically regarded as a part of Europe despite their geographical position just off the coast of Africa) and is part of a UNESCO Biosphere Reserve. The archipelago is now the largest marine protected area in Europe and has been proposed as a new Spanish National Park. These designations will go a long way to protecting the range of wildlife for which the archipelago is justly famous. Of course they are much too late to save a species like the black oystercatcher, although its demise was not the only one at the hand of collectors. On Montaña Clara in 1913, Bannerman found a party of four chats which were unfamiliar to him. He shot two and the others fled, perhaps understandably, and were never seen again. The chat was a previously unrecorded bird *(Saxicola dacotiae murielae)*, related to the rare Fuerteventura chat, and presumably restricted to these small offshore islets. It is speculated that the Chinijo chat may have been in terminal decline in any case because of the degraded habitats in which it lived and that Bannerman's intervention merely helped it on its way.

The islands in the Chinijo Archipelago are fascinating and exciting places, so far off the beaten track and yet, in other respects, so close to burgeoning human activity. Is it too selfish to tease the mind and shed a tear for modern Graciosa? In memory it is a wild and lonely place, a desert oasis, with a tiny community of tough and friendly fisher folk and wide dusty horizons. Today that image has been overtaken by the insatiable demands of tourism. Graciosa now has several hundred inhabitants, a school, lyceum, two supermarkets, a post office, bank, even an ATM, and, inevitably, tourist accommodation. The little harbour where the fishing boats were daily pulled up on the beach, now boasts a marina with lines of yachts. Nonetheless, the limit on development has been drawn at where it is now and the remainder of the island is a statutory nature reserve within the Chinijo Natural Park. Nonetheless, the beaten up Landrovers and hire bikes around the island are doubtless not going to leave.

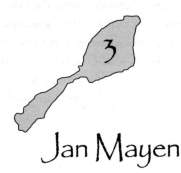

Jan Mayen

Beauty, it is said, is in the eye of the beholder, but on the remote island of Jan Mayen the beholders fall into two distinct camps. Idealists see it as a land of stark, primal beauty—the raw materials of the earth's creation—while realists see little more beauty in it than in the waste clinker and ash heaps from a furnace—which, as it happens, is precisely what it is.

It must be the ultimate monochrome island—white and black. Its origins are volcanic, for it was thrust up out of the sea, eons ago, and is the most northerly of all the islands that have their creation in the explosions of the Mid-Atlantic Ridge, the separation zone of the European and American tectonic plates. The volcano, Beerenberg, occupying the northern end of the island, rises to an impressive 7,470 ft (2,277 m). It was generally assumed to be inactive until it unexpectedly erupted in 1970, spewing lava into the sea and eventually adding a little over 1 square mile (3 km²) to the island's surface area. The mountain wears a permanent mantle of snow and a large ice cap, which spawns twenty separate out-run glaciers, and the whole island is habitually wrapped in an overbearing shroud of grey cloud and fog. Away from the glaciers and snow covered areas, the landscape is a mixture of tortured, twisted black lava cliffs, and coastal flats of equally black lava sands and pebbles, lapped by a narrow margin of white sea foam. Only in the short months of summer does a veil of deep bright-green mosses briefly relieve the landscape, in guano-rich slopes below the bird cliffs.

Approaching by sea from the east, the volcano is invariably lost in its all-enveloping foggy shroud and the outline of the long south-west pan-handle of the remainder of the island looms dark and unwelcoming below a menacing canopy of cloud. There is no such thing as a harbour on Jan Mayen—nor even the most rudimentary jetty—and on those rare days when a landing is possible, it is achieved on a beach at Kvalrossbukta on the western side or, more usually, at a narrow isolated beach on the east. Here, landing is a complex and singular operation. On such a day one September, the sea was choppy; too choppy for a zodiac to put us dry shod on the land. Therefore as the zodiac crested the swell towards the shore, a large lighter was tractor-shunted into the sea on its heavy metal cradle and the slightly chancy transfer was made from the zodiac to the cradle with the help of a short ladder. Once the lighter was hauled ashore a more substantial ladder assisted the long drop to the beach. No bush or shrub can tolerate the cheerless weather conditions on Jan Mayen but the beach here as elsewhere around the island, and indeed on so many Arctic strands, is littered with logs of long-lasting, sea-washed timbers from distant Siberian rivers. Among the logs there is the first evidence of former human activity: the bleached vertebrae, skulls, and other bones of whales. It is an Arctic charnel-house, testifying to the wholesale slaughter of the huge creatures that took place here in previous centuries. It is not only the passing of whales that is brought to mind here, but also that of some of the men whose purpose it was to hunt the great animals. Nearby, and juxtaposed to the remains of the whales' bones, are the resting places of some of the whalers who died on the island. Two memorial crosses and a memorial stone add to the funebrial atmosphere. This landing beach is ringed and overhung by dark cliffs, producing a grim, claustrophobic atmosphere, emphasized by the low ceiling of dark cloud. There is only one route out of the enclosed beach, leading steeply up the south side and passing through a wide lava gully at the crest. Quite suddenly the vision changes, leaving behind the dark, enclosed landing and breaking out into a wide panorama across the low ground, black sands and lagoons, stretching far down the east side of the island.

The island was first discovered in 1614, by the Dutch, although it is likely that the Vikings and possibly the Irish voyagers already knew of its existence and Henry Hudson probably saw it in 1608 on his journey to the north-west. In any event, once the island was known and the abundance of cetaceans in the surrounding seas realized, whaling very soon followed. Dutch whalers were the first people ever to live on the island, albeit seasonally. By 1616 there were 200 men working in six temporary processing stations, where rough shelters of canvas sails were gradually replaced by more substantial wooden buildings. For a short period of three years between 1616 and 1619, there were as many as twelve ships working the island and at that time there must have been considerably more than 200 men in total, some on land and some on the ships. On these beaches in the seventeenth century Dutch whalers made their summer camps and looted the surrounding seas of the great whales, especially the mighty bowhead (the Greenland whale) which occurred in large numbers. This Arctic whale spends its whole life on the fringes of the ebbing and retreating pack ice. It is the second heaviest of all the great whales after the blue whale, weighing up to 70 tons or more and has a massive bony skull for breaking through ice as thick as 18 inches (45 cm). Its 2 ft (60 cm) of blubber is the thickest of any marine creature. Unlike other baleen whales which feed on concentrated shoals of prey, the bowhead swims forward with open mouth (the largest in the world) continually filtering the intake of water, for copepods and other small marine creatures,

FIGURE 3 Eighteenth-century engraving of Jan Mayen.
© Courtesy of the New Bedford Whaling Museum.

through its baleen plates. They are very long lived, maybe the oldest creatures in the oceans with some individuals suggested as reaching 150–200 years old.

The whalers valued bowheads more than any other whale as a large specimen could yield as much as 25 tons of oil, rendered down from its blubber and a ton of whalebone (baleen). It had the additional advantage that, having a huge covering of blubber, it floated when killed, rather than sinking. It was also a very slow mover and therefore easier to pursue than others such as the fin whale. The whalers also hunted the appropriately named (northern) right whale which, around Jan Mayen, was at the northern edge of a much wider range. Like the bowhead it floated when killed and in that respect was clearly the 'right whale' to hunt. In a mere thirty years at least a thousand of these two species were slaughtered and processed for their oil and baleen. When they were first discovered, these whales were abundant in the loose pack ice of the Greenland Sea around Jan

Mayen where harvesting of these two species by the Dutch was intensive while it lasted. However, the operations on Jan Mayen were short lived because in 1632 two Basque captains employed by a Danish merchant, in an act of industrial revenge for having been evicted from Svalbard, raided the island and sacked the Dutch stations together with all their equipment. The industry here was predictably foredoomed: by 1640 the whales had gone and so too had the whalers. Jan Mayen was again deserted. In the shameful story of the virtual eradication of these whales in the region, Jan Mayen represented the nearest and most accessible part of their High Arctic range. After the demise of whales here and in Svalbard waters, the whalers moved ever further west and north over the next two centuries to complete their sickening harvest. Before the onset of commercial whaling, it is calculated that the population of bowheads alone, in the immediate area between Svalbard and Greenland, was probably upwards of 80,000 individuals. Today the population is only a fraction of that.

The Dutch whalers found little growing on the island to help with subsistence. Vegetation is negligible and the lava-blackness of the surface is relieved only here and there by the mosses and lichens below the bird cliffs, where the ground is fertilized by the annual bounty of guano. Scurvy grass grows in similar situations and by the sixteenth century was certainly already known as a remedy for scurvy and doubtless used to advantage by the whalers. Otherwise the list of plants on Jan Mayen is modest. Several species of arctic saxifrages and at least three endemic dandelions prosper in sites that offer a little more fertility and shelter from the maritime climate, high humidity, fierce winds, and lack of sunshine. The delicate glacier buttercup with its deeply segmented leaves and shy yellow-hearted cream flowers is the most beautiful of the flowering plants.

The inhospitable land surface may support little vegetation and no land mammals—polar bears are rare winter visitors and Arctic foxes, once numerous, are now extinct—but in spring the jagged cliffs are home to thousands of breeding seabirds. The bird community here, at 71°N, is typical of other similar places in the Atlantic section of the Arctic Ocean. There are Brunnich's guillemots lining the lava ledges, parties of little auks swirling high above the cliffs, colonies of colourful puffins, noisy penthouses of kittiwakes, and inevitable pairs of rapacious glaucous gulls, standing sentinel, awaiting the chance to plunder waifs or snatch the youngsters from others' nests. On the shoreline rocks there are groups of black guillemots, black as printers' ink and matching the rocks, given away only by the bold blaze of white on their summer wings and their whimsical scarlet feet. Above the tideline and around a long lagoon on the narrowest neck of the island (a second lagoon is virtually dry in summer) small colonies of Arctic terns have their nests and resolutely harass any intruders approaching their territories. There are pairs of ringed plover too, long distance travellers who cross the ocean to find sanctuary for breeding on Jan Mayen's lonely beaches. Purple sandpipers too are busy on the tideline. They are one of the hardiest of all northern shorebirds, some individuals not migrating south with the rest, but wintering it out on whatever Arctic shores they can find that are ice-free through the coldest months. Common eiders, the most abundant sea duck in the world, are tough denizens of these northern waters too and in late summer large parties, often comprising several families combined, gather offshore, rising on the swell and immediately disappearing in the troughs.

However, one species outnumbers all the others combined and is the pulse of the island's life: the fulmar. An approach to the island over the sea begins to hint at the numbers to be found. Fulmars in their hundreds are all around the ship, scudding low over the white-capped swell, lifting effortlessly into the wind or riding high, almost in touching distance on the uplift along the ship's side. Closer and closer to the island the picture is still more crowded with thousands upon thousands of fulmars, filling the air. On the nesting cliffs, as the birds soar up by the ledges, there is an almost constant raucous cackling as the birds greet mates or impinge on neighbours' sites. However at sea, amongst the maelstrom of birds there is a silence which is almost uncanny, as though no swirling mass of this magnitude could be so silent. There are notable fulmar colonies in other parts of the North Atlantic and Arctic—St Kilda is a prime example—but nothing can rival the sheer numbers here: somewhere in the region of 160,000 pairs nest on the island's cliffs. Look inland and the cliffs are studded with the white dots of fulmars sitting on their nests. Some individual birds are clearly of the northern morph, so called 'blue fulmars'. Instead of the usual snow-white bodies and pearl wings these birds have entirely dark grey plumage, standing out from the rest, almost like a species apart. Jan Mayen has an unenviable climate, wet, windy, and foggy much of the time, but one redeeming feature is that its geographical location means that it does not suffer the extreme winter temperatures or frozen seas of its nearest neighbour, East Greenland, although pack ice periodically sweeps the island. For the fulmars this means that they have no need to escape to the south, but can stay in the vicinity year round and benefit from the continuous bounty of the sea. Jan Mayen is indisputably the island of fulmars.

After the whalers abandoned the island around 1640, it was visited only occasionally during the following 250 years, by curious visitors, trappers, scientists, or fishing vessels and had no permanent human occupation until 1921. In that year the Norwegian government opened a meteorological station and with skeleton manning the island has been inhabited year round, with only one interval, ever since. During the Second World War the weather station was a key one for the Allies and also regarded as such by the Germans who tried unsuccessfully on several occasions to establish a base there themselves. In 1960 a Long Range Navigation (LORCAN) station was also set up operating for both military and civilian purposes. A small permanent civilian force of some twenty people now remains on the island all year, under the command of the station head who is a military officer. Their only servicing is by an occasional plane landing on a gravel strip. The accommodation base on this bleak and windswept outpost has an almost fairytale aspect to it—like stepping from the Stygian blackness of a dark cavern into the dazzling brightness of day. The long, low buildings—white of course—are wonderfully equipped with every facility imaginable for ease of living in the isolation of the lonely months of the Arctic winter: central heating, thick carpets, television, luxury lounge, recreation facilities, etc. Outside they have even built their own rudimentary swimming pool!

Everything here really is overwhelmingly black or white, with the exception of the scattered patches of moss-green. On a heavily overcast autumn day I stood on a jagged lava stack by the landing beach, watching the moving clouds of fulmars over the sea. Among the throng of birds I sensed a different one in the sky above me, cruising round. I looked up to see a great white bird, circling widely, a gyr falcon from the summer cliffs of Greenland, on its way via Jan Mayen to Iceland perhaps, or even further south; a spectral ghost, echoing the lost souls of the past on this challenging land.

4

Mykines

If you approach Mykines by sea from the east, via Sorvag on the island of Vagar, the first of many dramatic sights you meet in the long fjord is the islet of Tindholm. It is an astonishing mountain-island that has apparently been surgically sheared clean in half. The Faeroe Islands are rich in myth and legend and the remarkable shape of Tindholm is attributed to the island being cleft by an infuriated giant, dumping the southern half into the fjord depths and leaving in its place a sheer, unclimbable 800 ft (244 m) precipice. The remaining half of the islet is certainly uninhabited—it could hardly be otherwise—although even so there are signs of attempted habitation sometime in the past. Legend conveniently lends support to this suggestion. A man and his wife lived there with their small child. While the man was out at sea one day an eagle came down and snatched the child, carrying it off to its lofty eyrie on the cliff top. For love of her child the woman managed to scale the mountain top only to find the child dead. To this day, one of the five small pinnacles on the summit is still called the Eagle's Rock. To add spice to the legend, Tindholm was in fact the last breeding place in the Faeroes of the mighty white-tailed sea eagle, before its extirpation at the hand of man, as a deemed threat to his flocks.

Beyond Tindholm the grey mass of Mykines lies under a misty cope which too frequently hides its higher parts. The passage along the south side of Mykines passes close under gigantic cliffs, with giddyingly steep grass bluffs and outcrops with innumerable ledges, studded with the white chests of standing puffins and nesting fulmars. Interspersed with these outcrops is a series of perpendicular west-facing cliffs each one smothered with breeding kittiwakes, thousands of them swinging back and forth like white flakes in front of the black cliffs and thousands more perched on flimsy nests glued to impossibly tiny ledges. The whole way, over the heaving sea, there is a constant flow of birds, kittiwakes, puffins, and fulmars but also shags, guillemots, Arctic skuas, and occasional gannets. Without actually experiencing this maelstrom of birds, it is difficult to grasp just how enormous the numbers of sea-fowl are on this distant island.

There is no harbour on Mykines, only a landing place towards the western end, in a narrow gully open to the south-west, overlooked from overhead by yet another colony of noisy kittiwakes. A ramp leads from the sea up the side of the rocks, down which a cradle is lowered and onto which island boats are drawn and then winched up daily to the safety of an upper platform; no boat would be safe left in the recess of the gully overnight.

Mykines is the most westerly of the Faeroe Islands, a finger of land pointing out to the west, in the face of the worst weather that the North Atlantic can throw at it. And it can be very severe. Where a reading of 12 on the Beaufort scale signifies a hurricane force wind, storms of up to force 14 have been recorded at the lighthouse on the western point. One can only imagine the tumult of the sea at such a time. The island is about five miles (8 km) from east to west and little more than half a mile (800 m) at its widest. It is completely cliff bound and nowhere on the island is there safe access to sea level. Mykines is actually two islands, for at the western end, beyond the landing and reached, improbably, by a narrow steel bridge over a gaping chasm 130 ft (40 m) above the sea, is Mykines Holm.

The Holm is a small entity, no more than a third of a mile (500 m) long, terminating at the lighthouse which can only be reached because of the bridge. The islet is wedge-shaped, with beetling cliffs on its north side, sloping down to a wild rocky foreshore on the south. At the far end stand two formidable, bird-filled sea stacks, Pikarsdrangur and Flatidrangur, whose names neatly define their different profiles. Mykines village is strategically sited in a sheltered saddle of land on the main island, above the landing, alongside the island's one significant stream, with the in-bye land around it and the rising uplands occupying the bulk of the island away to the east. The path to these wide moorlands climbs from the village and passes a small lake which is white with hundreds of bathing kittiwakes. Like other gulls they will forswear salt water and always choose to bathe in freshwater if they can find it. The quiet moorland pastures are a peaceful world of difference from the endless hubbub of the pulsating bird cliffs all round the coasts. Here the gentle parachuting song flight of meadow pipits and the music of skylarks high in the sky are the sounds of the open hill. Widely scattered pairs of Arctic skuas have their nests on the moorland and leave the ground to threaten and harass intruders in unnervingly silent aerial attacks. Towards the eastern cliffs a colony of Arctic terns occupies the great bowl of Kalvadalur and mountain hares, introduced years ago, are the only mammals on the hills other than a scatter of sheep.

There have been people living on this island for at least 2,000 years. The earliest occupants were presumably Norse settlers although, as in innumerable places round the Atlantic coasts of northern Europe, it may well have been the seemingly ubiquitous Irish monks who first landed there; certainly the origins of the name 'Mykines' are claimed to be from Celtic roots rather than Norse ones. Is this the island that the famed voyaging monk, St Brendan, called the 'paradise of birds' in the sixth century? If so, the name could hardly have been more appropriate. Was it the Irish monks who cultivated the island, and were responsible for introducing the house mouse as long ago as the sixth century? In its island isolation, the mouse has evolved with time into an identifiably separate species, the Mykines house mouse. It is a burlier and brighter animal than the conventional house mouse, but its adjectival name is in reality a misnomer, as here it is predominantly an inhabitant of the bird cliffs, rather than the houses (an interesting contrast with the St Kilda house mouse (p.133)), although it is certainly present there too. As the mouse is principally an inhabitant of the bird cliffs, it is a passing thought as to whether there could be a future problem along the lines of that on Gough Island in the Tristan Da Cunha group (Chapter 11) where 'super mice' have been eating sea-bird nestlings alive and devastating the populations.

The modern village on Mykines proclaims its Scandinavian character with the forty stone and timber houses gaily painted a riot of white, red, yellow, or blue. The oldest ones, including the little church at the top of the village, are turf roofed, on which the grass is mown when necessary by the simple expedient of introducing a sheep. In common with the trend on so many island outposts, the year-round population of those who are permanently resident on Mykines has fallen to no more than a dozen. In the years between the two World Wars as many as 170 islanders lived there all year round. They were—as the current people are—subsistence farmers, fishermen, and efficient harvesters of seabirds. Now, the majority leave for the main Faeroe Islands as autumn tightens its grip, only to return to the island for the summer months. The weather on the Faeroes may be some of the worst on any of the island groups in the North Atlantic, with frequent strong winds, rain, and fogs. However, given sunshine days in June and July, Mykines is transformed into an idyll, its hay meadows bright with buttercups, red campion, angelica, sorrel, and marsh marigolds, puffins polka-dotting the grassy slopes and the skies busy with the comings and goings of innumerable seabirds. On days such as these the whole island basks in vibrant colours, warm sunshine, and a feeling of well-being.

Throughout the Faeroes, from time immemorial, the islanders have harvested seabirds as a vital and sustainable source of food. Apart from its imperative need, the tradition is a deep-rooted element of the Faeroese culture, to the same degree as the opportunistic communal slaughter of long-finned pilot whales. They are both traditions that help to define the Faeroese people. Throughout the islands a variety of seabird species is taken, largely depending on what is available on the individual islands and the relative ease of catching them. The most important species on Mykines is unquestionably the puffin. When the human population was at its highest, some 30,000 birds were taken in a season and even in recent years the catch has been similar. In years of particularly favourable weather conditions, twice or even three times that number might be caught. In an average day an experienced fowler can take 200–300 birds, although the records for a day's catch go well in advance of such numbers. In some of the early years of the present century puffin numbers and breeding success have been very poor and the catch consequently below average. Many of the birds caught are eaten fresh—two are considered a usual amount for one person's meal— together with potatoes and rhubarb jam (rhubarb grows prodigiously on the island) but the majority are salted down, or in recent years frozen, to help through the lean months of winter. The meat is rich and tender (and even better smoked!) with an understandable essence of fish!

Traditionally the puffins taken in summer are mainly caught by a method unique to the Faeroes, using a *fleyg*. This is a triangular net on a heavy ten foot pole which is wielded requiring considerable strength and dexterity. Each fowler operates from his own traditional and exclusive position on the cliffs, the same small hollow being used every year as it has been for generations of fowlers past. The hunter sits out of sight with his net flat on the ground beside him as the birds circle round in droves like locusts in front of the cliff. When an individual bird is identified as a target, the net is swung high into its path, the entangled bird being brought down, its neck snapped, and the corpse tucked into a belt encircling the fowler's waist. He eventually walks home with a full girdle of puffins. The *fleyging* is tightly regulated and over the centuries, despite the heavy annual toll of birds taken, the custom has never been responsible for a reduction in the size of the breeding colonies. The traditional starting date for *fleyging* is 2 July—a date as significant to the fowlers as 12 August is to grouse sportsmen in Britain. By this date breeding puffins are feeding young in their burrows and thousands of non-breeding birds begin to visit the colonies, circling round in myriads in the evening flypasts, so that, from a distance, the sky above the vast colonies on Lambi and Dalid appear to be under an uncountable cloud of midges. The fowlers claim to be able to recognize the difference between breeders and non-breeders but, whatever the truth of the claim, they will certainly never knowingly take a bird that is carrying a bill full of shining sandeels destined for a chick deep in the home burrow.

However, in the early years of the twenty-first century, in common with many other North Atlantic colonies, the Faeroes have seen the puffins suffer serious declines in numbers. In some seasons food at sea has been so short that virtually no young have flown from the colonies. By 2011, as in previous years, breeding seasons had been so bad that few, if any, young puffins were produced on Mykines or on other Faeroe islands. As a result, for the first time in history, the human harvest of birds did contribute to the depletion of the colonies. The puffin is not legally protected in the Faeroes although there are attempts to have a moratorium on catching while the population levels remains critical. This problem of declining populations is not restricted to puffins, but also affects Arctic terns and kittiwakes, both of which species have suffered successive bad breeding seasons and negligible recruitment into the breeding populations.

However, the other main species that is traditionally taken in numbers on Mykines is the gannet. The island boasts the only gannet nesting colony in the Faeroes with a modest population of some 2,000 pairs, increasing slowly. The colony occupies one wide ledge along the whole of the northern cliffs of the Holm, with outliers on Flatidrangur and Pikarsdrangur. Thus the entire Faeroese population is on this one small island.

The Holm is a quite extraordinary place for breeding birds. For its small size—600 by 200 yards (550 m × 180 m)—the density of birds makes it one of the most crowded sites anywhere. The number of breeding species alone is impressive. There are puffins in great numbers, gannets, Arctic terns (a huge colony, although it has suffered a major decline in recent decades), shags, kittiwakes, fulmars, guillemots, razorbills, black guillemots, oystercatchers, and eider ducks. All through the summer nights the island still echoes to the paranoid screaming of Arctic terns, but that is only the overture to a bewitching chorus that starts as the light fades and darkness sweeps the island. It begins with the soft purring of thousands of storm petrels amongst the rocks and in their earthy burrows, as the sitting birds call to attract their home-coming mates. The night time chorus is soon strengthened by the spine-chilling witch's laughter of the slightly larger Leach's petrels as they too make land and return to their burrows in the soft turf. The storm petrels are black plumaged, the Leach's only slightly paler, and the air is thick with them, but even against the dark sky their flashing outlines can be made out as they career around the breeding slopes and the white rumps of both species are occasionally glimpsed in the dark. There are yet more voices to join the nocturnal choir. In the depth of the dark, especially on moonless nights, Manx shearwaters add their wild calls to the voices of the petrels, as they too come home from the sea, flying round unseen and landing with a thump before disappearing into their underground tunnels. The Holm is as densely populated by night visitors as it is by the diurnal species. Unseen and unimagined during day time, the whole surface of the island is a honeycomb of nesting burrows.

On 25 January each year the islanders ceremonially troop to the cliff edge that overlooks the Holm to welcome home the returning gannets, for this is the day, according to tradition, for their reappearance. However, it is eight months later when the islanders come to the Holm itself for a more sinister reason, to take their harvest of the young birds when they are at their fattest, just before they are ready to fly. The northern cliffs on the Holm are over 425 ft (130 m) high and the descent to the breeding ledge is not for the faint hearted, as I can testify. Many years ago, Leon Heinesen, who was lighthouse keeper at that time, demonstrated the simple technique. A strong broom handle is pushed at an angle deep into the turf on the cliff top and a long coil of rope wound round it twice at ground level and looped with a half hitch. The remainder of the rope is then thrown down the cliff, down which the fowlers descend, reaching the broad shelf where the gannets nest. Alarmingly, this hair-raising operation is carried out in the darkness of night (cf St Kilda, Chapter 13), while the gannets sleep. The fowlers, working in pairs, abseil down to the ledge, untie from their rope and move quietly among the gannets, which are conveniently heavy sleepers, killing the sentinel and wringing the necks of the young and leaving the bodies where they are. After daybreak the corpses are thrown down into the sea for the waiting boats to collect. The men come to take the young gannets as weather permits, at the end of August or in the first days of September. In 2008 540 young birds were collected in this way and, in 2009, 570 plus a further 90 from Flatidrangur where the ascent of the stack is made from boats. Traditionally half a dozen men climb the stack and surround the young birds, corralling them into the middle of the flat top where they can easily cull the number they want. At one time small numbers of adult birds were taken in spring but this has now ceased.

The other wild meat which the Faeroese relish, despite its unpredictable occurrence, is whale. Deeply embedded in the Faeroese psyche is the hunting of schools of long-finned pilot whales ('caaing whales') which are likely to appear anywhere around the islands, during the months of summer or autumn. These are small whales, between 12 and 20 feet (3.5–6 m) long which occur in schools, running sometimes into hundreds. They have been exploited in these islands far back into antiquity and no opportunity is ever missed to take advantage of their appearance. This is the *grind*, the hunting and slaughter of entire schools of pilot whales, providing a bonanza of meat for a large number of people. Once a school is sighted near land, the word is passed and every available boat is put to sea. When sufficient boats are gathered, the school is gradually herded landward until the whales are eventually driven inshore on a beach or into a harbour, locked in by the encircling boats. The ritual slaughter that follows is not for the squeamish, for the panicking animals, thrashing wildly in the shallow waters, are slaughtered with lances and knives while the sea literally runs deep red with their blood. The spoils are shared out among the islanders, to time-honoured laws and customs, as an integral element of this ancient economic and social event. This practice is a perfect example of the widespread clash that occurs around the world, particularly in the context of aboriginal or native groups, between the claim that the practice has been a central part of the culture from time immemorial, while on the other side the claimed need for such killing to provide food in the modern world, is vigorously contested. This conflict is certainly alive in respect of the Faeroese *grindarǎp*. Every year several slaughters may take place. In a recent example it is claimed that, in July 2010, at least 236 manually slaughtered pilot whales were lined up on the dock at Klaksvik (Borðoy Island). The practice is overseen by the Faeroese authorities but not approved by the International Whaling Commission. It is universally denounced outside the Faeroes but the ears are deaf and the ritual killing continues unabated.

On cliff-girt Mykines no beach exists onto which a school of whales can be driven ashore and captured. The only source of whale meat the islanders have is when whales are found on the neighbouring island of Vagar, herded into Sorvag fjord and then beached on Bour or on the north shore of Tindholm. I have tasted whale meat thus obtained, not fresh, but air dried and found it distinctly unappetising, rather akin to a hybrid between putrefying mutton and a section of rubber tyre. However it is greatly valued throughout the islands and in centuries past was a crucial food in the lean months of winter.

The full-time resident population of Mykines may have fallen in recent decades, but the island will survive, of that there is no question. There is too much tradition and commitment to the island life there for it to disappear, go the way of St Kilda and Mingulay (Chapters 13 and 9) or become only a summer home. Anyway, there are cattle, sheep, and ponies to care for through the winter as well as poultry. Moreover, tourists have now discovered the island and there is ready accommodation there. You can even arrive by helicopter if you can afford it. Its location far out in the Atlantic and its wealth of birdlife are a strong draw for visitors who find that the island exudes a powerful atmosphere of real vitality and vibrant life which echoes to deep roots of history and a unique, sea-born culture.

Guam

Guam is—or was—an oceanic island completely clothed in a dense and vibrant canopy of tropical forest. All forests, not least those on isolated islands, are complex systems, for there is a delicate and symbiotic interdependence across all the different taxa within the forest. Such systems, individually unique to each island, have slowly evolved over the inexorable passage of time. Guam was no exception, but the balance there has been so severely upset that the island is now little more than a complete biological disaster. No other island, anywhere, has had its flora so damaged and its native birds and mammals so comprehensively eliminated.

Set in the tropical Pacific, some 13° north of the equator and at the southern end of the Mariana Islands, Guam lies in a coral sea 1400 miles (2,250 km) east of the Philippines, with Hawaii a further 2800 miles (4,500 km) away to the north-east. There is no doubt that, in its original state, it was a tropical Utopia, its natural forest cloaking the island with a canopy of green vegetation as the building block of the rich and diverse ecosystem within it. The forest resounded with the songs of birds and the hum of insects and the coasts were alive with herons, turtles, and migrating shorebirds. Nine species of native geckos and skinks inhabited the island (and still do), together with a monitor lizard, which may questionably be native to the island. The only indigenous mammals that have ever been present on Guam are fruit bats. Like other oceanic islands, Guam slowly developed its own flora and fauna, evolving a wide spectrum of tropical wildlife within which was a range of endemic species so that, like so many other remote Pacific islands, it was, in every way, a pristine Eden.

The picture today is one of the catastrophic change. Human encroachment has been all-pervading, removing much of the original forest cover, leaving the remainder in isolated areas restricted to the north and south ends of the island. Two enormous military bases occupy 30 per cent of the land and a major tourist hub dominates the coastal site at Tumon Bay on the west side of the island. Sadly, even the two remaining areas of forest have not escaped serious adaptation. In the north the non-native Tangan Tangan bush was introduced in the late 1940s as a fast growing replacement for the serious damage that had been wrought to the forests during the Second World War. The alien plant has prospered and readily outcompetes much of the native forest growth. At the other end of the island, large areas were clear felled during the Japanese wartime occupation and converted to agricultural land to provide food for their troops. Subsequently these areas were abandoned and have developed as extensive savannas of non-native grasses.

Notwithstanding these modern impacts, man's influence on the island's environment and many of its individual species reaches back over the centuries. The earliest people to arrive came from Indonesia, possibly as early as 2000 BC. At some stage in the past it was probably they who accidentally introduced Polynesian rats, the first unintended mammal to arrive on the island and one which, like others of its kind, is a well recognized predator on birds and their nests. Until this time birds, and particularly flightless ground-living species such as the Guam rail, knew no natural predators. The arrival of European man in the seventeenth century exacerbated the problem as he inevitably brought both the black rat and subsequently the brown rat as stowaways. The legacy of these introductions is that rats of one species or another now occur commonly especially in areas of human settlement and have put pressure on native bird species ever since, as well as becoming serious domestic and agricultural pests.

Spanish colonization of Guam began in 1668 and lasted until 1815. It was they who were responsible for several intentional introductions which have subsequently proved to cause lasting and continuing environmental damage. The pigs they brought with them from the Philippines have given rise to free ranging wild herds which occur throughout the forested areas of the island and are responsible for extensive damage to ground vegetation. The frequency with which one encounters areas of their rooting in the forest and its fringes gives an indication of just how many there are and the effect they continue to have. But pigs were not the only disastrous grazing animals that the Spaniards introduced, that have subsequently been left to run wild. They also brought Philippine sambar deer with them which have been used for hunting in the centuries ever since. Their numbers readily took hold in their new environment and they are found throughout the forested areas and the savannas where they stand accused of having a major impact on native plant communities. In the north of the island, within the National Wildlife Refuge, where pioneering tree planting of indigenous species has been undertaken in an attempt to replace the Tangan Tangan trees, a combination of deer grazing and pig rooting is recognized as a serious inhibiting problem. The deer are popularly hunted and are indirectly responsible for illegal wildfires that regularly blight many areas. Many fires are deliberately started by hunters to produce fresh growth of grass and thereby lure the deer into areas where they can more easily be shot.

Water buffalo (Carabao) too, came with the Spaniards who used them as beasts of burden, draft animals, and a form of transportation.

The animals had long been domesticated on the Philippines, certainly since pre-Hispanic colonization, but on Guam they eventually outlived their usefulness and were left to run wild. Up to the early twentieth century, astonishingly, there were many thousands of the animals in the southern half of the island. The herds were so numerous and the damage they were doing was so widespread, that various campaigns were undertaken to reduce the numbers, including unsuccessful attempts at birth control. By then, however, although the damage was done, the animals had become part of popular culture and regarded by many as a national symbol, with the result that elimination was popularly ruled out. The carabao today are confined to one area of navy property and although the environmental harm they continue to cause is substantial, it is at least fairly restricted.

Although the Spanish were also responsible for the first introduction of dogs and chickens, the blight of introduced creatures did not end with them but the alien cataclysm continued. In 1937 cane toads were brought onto Guam and have prospered in much the same way as they did when introduced so disastrously in Australia. The toads are voracious feeders, eating almost anything they can find including plants, insects, other invertebrates, amphibians, and even carrion. They are highly toxic, as are their tadpoles and they breed prolifically and have spread throughout Guam, where they invariably impact on other species. As yet they are not recognized as a serious pest although once introduced they are virtually impossible to eliminate. The Japanese chose to introduce giant African landsnails during their brief tenure of the island in the Second World War, which was a strange decision as the snail is recognized elsewhere as a serious agricultural pest and a species which most would regard as unwelcome almost anywhere. In an attempt to establish game birds for recreational shooting purposes between the 1950s and 1970s, the US government introduced the attractive ground-nesting black francolin among other gallinaceous birds. The francolin is the only one to have multiplied successfully and is now a common species outside the urban areas.

But still the list of harmful introduced species goes on. Asian shrews were found to be present in 1962 although it is not known when or how they arrived. Several non-native lizards, geckos, and skinks managed to find their way onto Guam sometime from the 1950s onwards. Although these animals might be presumed to be innocuous settlers, one of them in fact has an impact far beyond its size or status, as we shortly see. Furthermore in 2007 an infestation of coconut rhinoceros beetle was detected in the north of the island. The insect is highly damaging to coconut palms. Adult beetles fly at night and feed on the sap from fresh emerging fronds. If the growing tip of the palm is damaged in this way it can result in the death of the tree or at least serious leaf loss and consequent decreased nut production. Normally the beetles breed on the ground in decaying logs, but to great surprise, they have now been found breeding in the crowns of palms together with a self-contained community of crabs and other creatures, presumably benefiting from the fact that rats, which normally favour such sites, are now less plentiful. An aggressive programme of control has been initiated in an attempt to eliminate the insect.

The combination of extensive habitat loss and the impact of this whole raft of introduced species is enough to destroy the wildlife of any island the size of Guam. However, there is still one more factor which probably outweighs all the others combined, in the terminal effect it has had, principally on the bird population, but also on other species such as the fruit bats.

Sometime around the end of the Second World War, after the Americans had retaken the island following a bloody three-week battle, brown tree snakes were found to have made their way onto the island, probably by way of supplies brought in from further south in the Pacific. The brown tree snake, as its name suggests, is an arboreal animal, native to north-east Australia, Papua New Guinea, and its adjacent islands. In its home range it grows to a maximum length of 3–6 feet (1–2 m) and lives on a diet of birds, lizards, bats, and other small vertebrates. It is a mildly venomous colubrid with rear fangs that are grooved rather than hollow, well suited to disabling its normal small prey but not therefore regarded as a threat to adult humans, despite an aggressive, confrontational stance when threatened.

What it found on Guam was an abundant prey resource with seemingly limitless birds, lizards, geckos, rats, and domestic animals and no natural predators, except occasional monitor lizards. As a result its numbers burgeoned. It was not until the 1970s that people realized that something was seriously decimating the bird population on the island. There was clearly a catastrophic drop in the numbers of all eighteen native bird species, not only in inhabited areas of the island but also throughout the farthest corners of the forest. Reasons for the alarming declines were considered and dismissed— disease, pesticides, pollution, or perhaps rats—but even they could not reduce the numbers of all birds to the extent that was being found. It was only in 1983 that the reason eventually became apparent and the culprit was identified as the brown tree snake. By this time the snake's population had reached unprecedented levels and within years several native birds were on the brink of extinction. Endemic rails have occurred on many remote oceanic islands over time (where many were eliminated after the arrival of man). Guam is no exception and in the 1960s it was estimated that the unique and popular Guam rail (ko'ko') had a population of as many as 80,000 birds. By 1986 only nineteen individuals remained. The unique and beautiful Guam flycatcher has gone completely, exterminated together with the island's white-eye. The Micronesian kingfisher was reduced to no more than 18 individuals. The only forest birds left by the first decade of the twenty-first century were about 500 individual Micronesian starlings, some Guam swiftlets (see below), and a single Mariana crow. All other birds have been entirely extirpated from Guam. Nowhere else in the world has a snake been responsible for the extermination of birds in the way that has happened on Guam. It is estimated that throughout the island the tree snake has a population of some 20 individuals per acre (50 per hectare) with a likely total of a million snakes, making this the densest population of snakes anywhere in the world. As they can forage from ground level to tree top, nothing that represents potential prey is safe from them.

The situation created by the snakes is extremely disturbing in several respects. Not only has the bird community of the island been eliminated but, being strictly nocturnal, the snakes cause traumatic problems when they enter houses at night, even occasionally attacking sleeping babies and young children. Moreover, because it is arboreal it can climb virtually anything and anywhere, one result of which is that it has regularly crossed electrical transmission lines and caused extensive power failures. In the past some of these failures have been serious and have covered wide areas, including military bases. Because of the problem caused by the snakes, the whole electricity supply system throughout the island has had to be modified and adapted; sub-stations are now proofed against access by the snakes so that when power surges do occur, they affect much smaller areas.

Colourful and dramatic accounts have sometimes been written about the numbers of snakes on Guam and the scale of the problems they produce. However, despite their numbers, they are not easily seen everywhere, as popularly imagined, hanging off trees or fences like strings of spaghetti. They are strictly nocturnal and many people on the island never or only rarely see one even though they are so numerous. Certainly they occur around human settlements and in these places find an inexhaustible supply of warm blooded prey. Apart from rats, there are house shrews, mice, domestic chickens (or at least their eggs and chicks), and they are quite capable of taking small pups or kittens. One result of this super abundant food supply is that some of the snakes grow to unprecedented sizes; instead of a normal average of about 5 feet (1.5 m), they have been shown to grow to as much as 10 feet (3 m) in length! In the forest areas the situation is very different. Here there are essentially no warm blooded prey animals remaining and the snakes are forced to feed on skinks and lizards. Central to this is the strangely named curious skink, an alien species originating from Papua New Guinea which was first found on the island in the 1960s. Like the tree snake, it found conditions ideal and its numbers have boomed. It is estimated that there are between 4,000 and 8,000 skinks per acre (10,000 and 20,000 per hectare) throughout the island! With no alternatives, they are virtually the exclusive prey for the snakes in the forest areas, with the interesting result that juvenile snakes, with no warm blooded animals to balance their diet, do not thrive but grow to a certain length, mate, produce eggs and then die before reaching maturity. The vast population of curious skinks, together with other imports including the house gecko is entirely responsible for sustaining the tree snake in its forest habitats.

Predictably it is an ultimate ambition to see the snake eliminated completely from the island and, although impressive control measures are already taking place, this long-term aim is still a fairly distant dream. Various control methods have been tried, but by far the most successful to date is the use of large numbers of funnel traps, baited with a live mouse in a separate enclosure at the far end of each trap. These traps prove to be 100 per cent more successful than other methods. Effective though this is at a local scale and in seriously vulnerable locations, it is acknowledged that this technology is never within a whisker of solving the problem island-wide. Work is under way to try to develop a system of aerial delivery of toxic baits but it is fraught with difficult challenges, not least of which is that the baits need to be caught up in the forest canopy to be available to the arboreal snakes but out of reach of other creatures on the ground. Almost uniquely the tree snake also feeds on carrion. On the basis of this knowledge, it has been shown that a fresh mouse corpse, laced with 8omg of paracetamol (the active ingredient is acetaminophen) is guaranteed to kill the snake and is likely to be the preferred toxin in aerial baiting. The challenge is not only to develop the method of delivery, but to find a means of preparing thousands of baits—probably eventually glued to cardboard floats—that is not ridiculously demanding of man-power. In the meantime extensive efforts are made to ensure that the snake is not accidentally exported to other places in the way it arrived on Guam. Every aeroplane or sea vessel leaving the island is subject to a high profile search through the use of trained sniffer dogs. These have so far proved successful in intercepting occasional snakes that have found their way into departing cargo or luggage.

The devastation caused by the tree snake is resulting in several dynamic initiatives aimed at the restoration of some of the species which have been snatched from the brink of extinction. The most high profile bird on Guam is the rail—ko'ko' to the islanders. After the collapse of its numbers, the nineteen extant individuals were all caught and removed to zoos. These birds have formed the foundation of a captive breeding programme on Guam and all rails that exist today are descended from just ten individuals. Experience has shown that staff have to demonstrate a high degree of skill to achieve the necessary genetic matching of pairs to produce results. When this is successful the birds may start breeding at five months of age and can produce up to ten clutches of one to four eggs per year! The aim is to produce 100+ new birds each year but success depends as much on staff skill as on the birds themselves. Once the captive population on Guam reaches approximately 150 birds, fifty are sent to the nearby island of Rota for release. Initially there have also been three experimental releases on Guam as well as a first release of sixteen birds on the small rat-free Cocos Island off the southern end of Guam.

Another critical species being bred in captivity is the spectacularly colourful Micronesian kingfisher, the entire current population of which derives from only eighteen birds. In 1980 all the remaining birds were held in mainland zoos from where some returned to Guam in 2004 to enter the breeding centre. The birds have proved much more difficult to breed than the rail, sometimes exhibiting cannibalism among other problems. Recently, however, breeding has been successful and by early 2011 the population was 130 individuals. By that date, however, none had yet been released to the wild.

The Mariana crow is restricted to Guam and its neighbour Rota and although one pair is in captivity on Guam its future is indeed bleak. It is declining seriously on Rota, probably as a result of illegal shooting and numbers had fallen to 120 breeding adults in 2011. On Guam, where captive birds were released in a protected area in the north of the island, they disappeared some seven years after release, again probably due to unauthorized shooting.

Colourful posters are to be found around the occupied areas of Guam, urging people to help save the 'fanihi' and offering generous reward for information leading to the prosecution of anyone convicted of killing one. Fahini is the Mariana fruit bat. It is a fine looking animal with dark smoky fur, bright lively eyes, and a white chest band, but its future on Guam seems hopeless. There is only one specimen in captivity and allegedly only another seven in the wild, the remainder having been extirpated by the snake. No funds can be found to try to salvage the animal on Guam, part of the argument being that modest populations still exist elsewhere on the Marianas Islands.

So, the damage that has been wreaked by the brown tree snake is without precedent in its extent. Nowhere on Guam is the song of birds to be heard in the forest, thanks to the depredations of this snake. Away from the hubbub of the inhabited areas there is a claustrophobic silence that is total and eerie and is, for me, the haunting memory of the island. No bird flies across the road ahead of me, no overhead wire is the perch for birds and nowhere in the forest does any bird sing to welcome the dawn or lighten the day.

There is a cave in the south of the island where a small colony of endangered Guam swiftlets breeds. The snakes found the cave long ago and the number of birds in the colony fell to no more than 400 in the 1990s. The snakes were even found living in the roof of the cave, leaving only a handful of swiftlet nests that were out of their reach. Now a snake-free buffer zone has been created at the entrance of the cave. A significant achievement to celebrate is the fact that the swiftlets have now built up their numbers to probably a thousand or so.

What happens if one entire element of the forest ecosystem is removed, as is the case with birds on Guam? What part do they play in the balanced well-being of the forest? How long will it be before any effects are evident? The University of Washington has initiated a programme which is looking precisely at this issue. The project carries out detailed monitoring of seed deposition and insect activity in sections of forest with and without birds, using control sites on nearby Rota where the bird community of the forest is intact. Does the fact that butterflies are present in enormous numbers in and around the forest on Guam give a first hint of the effect of absent bird predators?

Guam is an extreme example—perhaps *the* extreme example—of the dreadful effects that man's careless management can have on the natural communities of an oceanic island. Modern Guam is defined by the silence of the birds.

6

San Blas Islands

The San Blas Islands lie just offshore from the Caribbean coast of Panama, some hundred miles or so east of the Canal, towards the Colombian border. The archipelago is a myriad of islands scattered like stardust along the Panamanian coastline. They are, of course, tropical islands, only 9° north of the equator and are hot and humid most of the year. With their low sea-level profiles they are dangerously pregnable to inundation, lying on the crystal surface of the Caribbean, no more than 3 or 4 feet (1–1.2 m) above the ocean's tideline, completely flat and unprotected from the sea. Fortunately they are tucked into a corner of the Caribbean where hurricanes are least frequent.

Most of the islands are tiny, frequently less than 2.4 acres (1 hectare) in size, and although there are almost 400 of them, fewer than fifty are inhabited. Those that are, are often packed with humanity. The universal *bohios* (palm-leaved, thatched huts) are cheek by jowl with each other, right to the water's edge where they are often linked to each other by narrow planks across inlets and creeks. The waterside is the centre of activity with a bustling confusion of small boats, dugouts, children swimming, fishermen sorting their catch, and women washing utensils or clothes.

The shallow seas surrounding many of the islands contain classic coral reefs—some of the oldest in the world—and the clear waters of the reef support kaleidoscopic colonies of corals, tropical fish, turtles, tunicates, sea urchins, sponges, crustaceans, and sharks; often in the shallowest parts, the underwater world of dazzling colour and movement is easily visible from the sides of a boat. The communities of fish are spectacularly colourful, with parrot fish, groupers, jackfish, damsel fish, clownfish, butterfly fish, and many others. Fishing, together with the harvest of coconuts, is a central element of the islands' subsistence and economy and many of the native fishermen still fish the offshore waters in dugout canoes (*cayuco*) in exactly the way that their forebears have done for timeless generations. However, even in this seeming paradise, all is not well. As elsewhere in the Caribbean, driven by increasing human numbers, man no longer lives in perfect harmony with the resources of the sea. Through overexploitation, many of the Caribbean populations of fish are now depleted, the two principal species of sea turtles which breed on these coasts, green turtle and hawksbill, are bordering on extinction and the reefs themselves are suffering serious damage from the effects of climate change, pollution and, in some places, increased sedimentation. Thankfully, although these are concerns on the San Blas Islands, the problems are not as serious around this area of the east Panamanian coast as in most other areas of the Caribbean. The autonomous San Blas islanders protect their marine environment jealously, for example by banning all scuba diving.

These scattered islands are tropical treasures, their skies graced by elegant and familiar birds that are characteristic of such places. The magnificent frigatebird does not breed here, but is common in the area, circling overhead on long narrow wings and deeply forked tail, with the consummate ease of the supreme flier that it is. It is a fisherman but also a pirate, robbing other birds such as the brown boobies and masked boobies, both of which are also numerous here. Everything about the boobies is sharp and angular—wings,

FIGURE 4   Dug-out canoe (cayuco).
© The author.

wedge-shaped tails, and long pointed bills—and they are the ultimate high divers, plunging Exocet-like from 40 or 50 feet (12–15 m)
into the water to take fish from well below the surface. Occasional
black vultures, numerous all over this part of the Americas, cruise
overhead, drifting out from the nearby mainland, silent, motionless
on the wing and vaguely disconcerting. The only gulls that occur
around the islands are laughing gulls, abundant winter visitors from
North America, here in springtime starting to make their way north.
Like the frigatebirds, they are not averse to robbery and will harass
the brown pelicans, even to the extreme of comically trying to perch
on the larger birds in an attempt to get them to disgorge their catch.

The sandy beaches on the quieter islands are used as resting and
refuelling stops by a great variety of shorebirds that coast along this
sea-board on their great seasonal movements between the two continents, although numbers are not as great here as on the Pacific
coast. In April the most evident wading birds are whimbrel, willets,

and semi-palmated plovers. The first two are long-distance travellers, heading for breeding areas far north in America and Canada. The willets are sturdy, dull grey-brown waders when seen on the shore-line, but are suddenly transformed in flight, showing bold, broad white wing stripes, black wing tips, and white rumps, announcing themselves with a strident 'kip,kip,kip' call.

I don't remember the name of the island I first landed on, one particularly bright April morning—nor indeed whether I ever knew it—but it was one of a scatter of virtually identical islets, dotted across the sea, palm-clothed and with strand lines of dazzling white sand. Apart from the fifty or so of the islands that are inhabited, all the others are owned and the valuable coconut palms growing on them are carefully guarded. Each island is a mirror image of the next, except that the one I was on clearly supported a populous and thriving local community. We drew our boat up alongside a rickety wooden jetty, crowded with beaming, excitable Indian children, some of whom, either by design or by being pushed, plunged down into the water and swam around us. The uncertain decking on the jetty trembled and moved as we walked on it, resulting in the whole edifice swaying disconcertingly. Apart from the worrying number of people on such an unstable structure, the only building it supported was a very small, wooden, long-drop dunny. It was far from private, gap-sided so that inquisitive children could readily smile at the occupant inside and offer helpful leaves through the gaps. As embarrassing, was the fact that the long drop into the sea was directly above the favourite place under the jetty where the other children splashed and wrestled.

At the head of the jetty, more of the people were gathered in front of the thatched-roof huts. These are the Cuna Indians, one of the purest aboriginal races now left on earth. The population of the islands is around 20,000 with the remainder of their tribe living in their original homeland along riverside villages in the mountains of eastern Panama.

The people are a fine looking race, friendly and happy folk, full of charm. They have their own language—Cuna—and are one of the best organized native groups in Central or South America. They maintain their old religion, closely related to their medicinal beliefs and serviced by seers and medicine-men and to date they have successfully resisted the blandishment of European missionaries. The San Blas Cuna gained autonomy from the Panamanian Government in the 1920s and have sovereignty over their affairs through an Islands Congress. They are a contented people of agriculturalists and fishermen. Most of the original vegetation of the islands has been long cleared and crops of yucca, maize, rice, and cocoa are grown as well as the ubiquitous coconuts. One of the problems is that there is no fresh water on the islands, leaving the inhabitants reliant on regular supplies from the mainland. Principally, however, the Cuna are fishermen, reaping a ready harvest of seafood from the shallow crystal waters and coral reefs around the islands.

They are tiny people, the second smallest in the world after the Pygmies and, over the generations, in the face of spreading westernization, they have assiduously protected their heritage, culture, and way of life. There is nowhere that I have ever been where there is such a sensation of stepping back in time when coming amongst them; there is a very special essence for, despite what goes on around them, they are still so far essentially untouched by the rest of the world. Although many of the men have adopted western shorts and T-shirts, the women are resplendent, in bright multi-coloured dresses and blouses and bedecked in traditional nose rings and pendulous ear rings, many of the older ones displaying wrists and ankles heavy with bangles of gold. They are amongst the most colourful of all native peoples and famous for the magnificent textiles that the women produce and sell in volumes to eager visitors. Examples of their work are hung outside all the huts along the paths among the palm trees. These craftswomen are world famous for the superb cotton 'molas' which they work in reverse appliqué and make into blouses and dresses, but they also produce glorious appliqué 'San Blas T-shirts' and similar with vibrant technicolour motifs of parrots, macaws, toucans, humming birds, turtles, and other creatures among which they live, emphasizing the bond with the wildlife that share the islands with them. All over the village, the women sit in the sunlight outside the huts working on the appliqués. Some grin broadly as we pass and point encouragingly to the textile masterpieces hanging behind them whilst some of the younger ones smile shyly and turn their eyes down.

7

Ascension Island

In the vastness of the oceans it is often the tell-tale drift of distant white clouds that gives the first hint of land on the far horizon. Certainly on Ascension the south-east trade winds gently brush the surface of the island and, as they rise, condense into pale clouds and exhale soft breezes on the summits, deceiving the upper slopes with an unfulfilled promise of rain. On closer approach, the island itself emerges from the dark green deeps on the Atlantic, not as a perfect volcanic cone like Pico or Tristan da Cunha, but broken-backed, barren, and rugged. It is a lonely milestone on the ancient sea routes for countless migrating marine creatures, a landmark for pioneering mariners and safe refuge for hosts of seabirds.

This small island was first seen by the Portuguese in 1501, although it was left to another Portuguese navigator, Afonso de Albuquerque, to record its existence on Ascension Day two years later, and bestow the permanent name on it. These early explorers, finding a barren lava-strewn island, populated only by millions of seabirds, would have difficulty in reconciling that image with the initial sight that greets a visitor nowadays. The first impression often appears unrealistic, for here in mid ocean is not a swirling mass of seabirds, but the modern face of our electronic and technological world.

Nonetheless there is still a real excitement about the island, a mere dot in the great expanse of the ocean, 1,000 miles (1,600 km) from everywhere but pulsing with human activity and, despite a legacy of 200 years of ravaged wildlife, still a precious treasury of rare seabirds, ancient reptiles, and endemic flora.

Remote in the deep Atlantic, Ascension has established its name as one of the world's vital communications hubs: a crucial military base in the Second World War and, in recent decades, the essential staging post for flights to and from the Falkland Islands and a key BBC relay station. This small basalt-lava island, primed with forty-four separate volcanic cones, is studded in parts with masts, antennas, satellite dishes, radars, and the equipment of a key listening station, all managed by complements of both military and civilian staff. The USA has had a presence here almost continuously since the Second World War and it was they who built the airfield and its enormous two-mile (3-km) runway in the south of the island on 'Wideawake Field', long enough to land the space shuttle if necessary.

Although Ascension was discovered in the great age of exploration, it was never settled, nor were there ever indigenous peoples living on this barren cinder bed of an island. It can claim no particular beauty, but speaks of primal tectonic forces and a savage, uncomplicated geology. In the seventeenth and eighteenth centuries the island was used as a victualling station by ships from many nations but, whereas the abundant seabirds and turtles could be plundered for meat, the virtual absence of fresh water limited its importance. It was not until 1815, when Napoleon was incarcerated on 'nearby' St Helena (in fact far away to the south-east), that a British garrison was billeted on Ascension to intercept any rescue attempt, thus marking the origins of the occupation that continues to this day.

The desert-like appearance of the island does not, however, mean that there is no native wildlife to be found. Far from it, because it is evident that, in the days of the early seafarers, the island supported vast colonies of breeding seabirds. It is estimated that the total may have been in the region of 20,000,000 pairs. It is equally clear that the use of the island by passing vessels was responsible for the disastrous but all too familiar invasion of ships' rats—that is, the black rat. As long ago as 1656, Peter Mundy, an English seaman, recorded the presence of a flightless rail (*Mundia elpenor*) on the island and has bequeathed us a rudimentary and intriguing sketch of it. The history of this interesting endemic bird is not recorded thereafter but it has evidently been long extinct: it is another in the long line of flightless birds on oceanic islands that vanished after the arrival of European voyagers and their rodent stowaways. The presence of rats on Ascension from at least the early eighteenth century heralded the rail's inevitable demise; if any individuals did manage to survive the rats, the arrival of cats with the English detachment in 1815 would certainly have been the final *coup de grace*. It is interesting that there is no evidence of any other endemic land birds on Ascension in historic times, although sub-fossil remains of an indigenous night heron have been found. This absence of other terrestrial species is perhaps surprising, bearing in mind that the more remote Tristan group of islands, farther south in the Atlantic support seven endemic species/ subspecies of land birds, together with two additional species that vanished on the main island of Tristan da Cunha itself, following human occupation, both of which were in all probability endemic.

Ascension lies about 8° south of the equator, in the path of the south-east trade winds which help to produce temperatures that, in tropical terms, are pleasantly tolerable all year. The seas around the island provide important feeding for seabirds, a fact which is reflected historically in the number of species and the large populations of many of them. The same species are still represented on the island today, albeit in vastly reduced numbers. Two members of the gannet family, brown booby, and masked booby are prominent, together with a smaller number of red-footed boobies. The spectacular red-billed and yellow-billed tropic birds with bright-coloured dagger bills and delicate, long white tail streamers, nest in hollows and crevices on the basalt cliffs. Both black and brown noddies, dark members of the tern family and the little Madeiran storm-petrel, together with numerous pairs of the delicately beautiful white tern (fairy tern) are also denizens of this isolated seabird fortress.

In addition to these more widely distributed species, there is one important endemic seabird on the island, the globally-threatened Ascension Island frigatebird. It is one of five species of frigatebirds found throughout the tropical seas around the world. At sea, away from colonies of individual frigatebird species, it is notoriously difficult to separate one species from another, as all are extremely similar in plumage and size. They are all large, light-bodied birds: spectacular fliers with long narrow wings and deeply forked tails, designed for effortless dynamic soaring. They can claim to be among the most skilful fliers of all birds for they can travel immense distances over the oceans with consummate ease but also have a devastating turn of speed and amazing aerial agility. They will plummet seawards in rapid pursuit of flying fish when the fish launch themselves above the sea surface to avoid marine predators, but they also feed on other small fish or carrion snatched from the surface. Ridiculously, all the frigatebird species, true seabirds though they are, can neither walk nor swim! Much of their food is gained in the manner described, but they are not averse to piracy and will relentlessly harass boobies and tropic birds as they return to the nesting colonies, forcing them to disgorge their catch. It is a habit that is reflected in their popular name and one which also earned them the old mariners' eponym—'man-of-war bird'.

The endemic Ascension Island frigatebird is wholly restricted to this one island and its surrounding seas, gracing its coastal skies with its soaring flight and distinctive silhouette. There are about 6,000 pairs breeding on Boatswainbird Island, just off the east coast of Ascension and it is the only frigatebird species found here. The small, sheer-sided Boatswainbird Island is important out of all proportion to its size. For this tiny islet has been the total salvation for

FIGURE 5 Bosunbird Island.
© Ascension Island Government.

Ascension's previous bounty of breeding seabirds. It is flat-topped and measures a mere 400 yards (365 m) in length. It is a white island—whitened through the deposits of guano deposited there by a thousand generations of breeding seabirds. Boatswainbird Island is, and always has been, rat-free. It is on this tiny islet that seabirds nest in dense numbers while on the main island the years have seen them inexorably reduced by the rats and cats until they have been almost completely exterminated. Ascension's millions have shrunk to the numbers that can be accommodated on the tiny Boatswain-bird islet. It remains, to this day, a remarkable sight, a living reminder of what Ascension has lost: an avian metropolis and an aerial blizzard of bird life. Here are all the nesting frigatebirds together with boobies, tropic birds ('boatswain birds'), and noddies. Even on this offshore stack, man visited his presence, not this time impinging on the birds themselves but profiting from their droppings. In the 1920s guano was harvested here for a time and even a very short railway built to move the cargo to the cliff edge for lowering to a boat below. Men actually lived on the rock in shifts for a few days at a time, serviced from the nearby beach.

However, despite this famous citadel with its cloak of birds, including the endemic frigatebirds, the species for which Ascension is probably best known is the sooty tern, locally known as the 'wideawake' on account of the ceaseless calling of birds in the colonies. It is a pan-global species which nests on many tropical islands around the world, often in huge colonies. This is the case on Ascension where the principal breeding site on the 'Wideawake Fairs', south of the runway and inshore from Mars Bay, provides an unforgettable spectacle. In the 1950s there were upwards of half a million pairs in this one colony but predation has resulted in serious declines over the intervening years. At present there are estimated to be in excess of 240,000 pairs and they are still a spectacular sight, with breeding birds scattered across the lunar desert, nesting in hollows among the jagged lava blocks. The salty air is alive with the comings and goings of thousands of birds; the sight in the 1950s must have been one to behold. They have survived in this colony purely because their numbers were so prodigious that the predators simply could not eat their way through them. They are interesting birds too, with an odd 9–10 month breeding cycle rather than the normal 12-month cycle of other tern species, or indeed most other seabirds. Away from the breeding grounds sooty terns are pelagic, spending the great part of their lives far out at sea. Once the young fledge they too move away from Ascension and spend their lives on the wing until, after six or seven years, they are ready to breed and return to their natal colony. Not all years provide good breeding results, for if there is a shortage of food at sea in a particular season, many of the birds may not breed or, if they do, they may be forced to abandon the eggs or chicks. Nonetheless, they are long lived birds and can afford occasional failures.

So, apart from the sooty tern colony at Mars Bay, all other seabirds have been confined to Boatswainbird Island and small offshore stacks with a few relict populations on inaccessible cliffs. However, their story has eventually changed for the better. In 2001, in an important initiative, the Ascension Island Conservation Department was established and, in conjunction with the Royal Society for the Protection of Birds (RSPB), launched a Seabird Restoration Project, funded by the UK Foreign and Commonwealth Office, aimed at the eradication of feral cats on the 35-square mile (91-km²) island and the serious reduction of rat numbers. Within twelve months of the initiation of the programme, the first evidence of recolonization was recorded with pairs of boobies nesting on flat land adjacent to the offshore colony on Boatswainbird Island. Steady progress has continued and, by 2011, monitoring showed that over 500 pairs of masked boobies and twenty pairs of brown boobies, together with almost 200 pairs of brown noddies and a handful of tropic birds were already breeding again on the main island. By that date none of the frigatebirds had yet returned to the main island to breed, although there were large numbers roosting. The project has clearly been an outstanding conservation success, a glowing example of what can be achieved, even in hostile conditions. Control of residual rat numbers continues and the project shines as a beacon of hope for the many other islands in the world where similar problems exist. Ascension was formally declared cat free in November 2006 although, in agreement with the community, domestic cats, duly neutered, can still be kept as family pets.

Away from the seabirds on the lower parts of the island, the highest crater on Ascension rises to almost 2,800 ft (859 m) on Green Mountain, which, as its name suggests, is the only area of the island that supports lush plant growth. The temperatures here are cooler, cloud frequently cloaking the mountain and ensuring a high level of humidity. Even here, there has been long-term grazing damage to the natural vegetation (the Portuguese introduced goats as early as the sixteenth century) compounded by introduced alien plant species, including conifers planted long ago to provide replacement ships' masts when needed. In parallel with the Seabird Restoration Project, the first decade of the twenty-first century also saw serious progress being made with the reduction of invasive plants and the commencement of a long-term project to cultivate and reintroduce endemic and native plants to the Green Mountain area. So serious has been the damage to the island's fragile native vegetation, that of ten known endemic plants, each is critically endangered, threatened or, in the case of four species, believed extinct. However, in a 2010—appropriately enough the International Year of Biodiversity—Conservation Department staff carrying out plant surveys on rugged cliffs on Green Mountain came across several tiny examples of the unique Ascension Island parsley fern. Once relatively common on the mountain, it was long believed to be extinct. Some of the sensitive spores were rapidly air-lifted to Kew, found to be viable and are now the basis of a regeneration bank of the plant. All the other native plants on Green Mountain are non-flowering species including other ferns, club mosses, and grasses. In 2005 Green Mountain was declared as Ascension's first national park.

The lava that flowed from the crater and its satellites dominates the whole landscape of the island and determines its rugged coastline. However, around its coast the jagged cliffs are broken in places and intersected here and there by sandy bays and coves. These beaches provide one of the most important breeding sites in the world for green turtles. A survey in 1998–99 (Godley *et al.* 2001) found an estimated total of almost 14,000 clutches laid in one season, with the peak of the nesting activity between New Year and the middle of May. The turtles have undoubtedly nested here since time immemorial and, prized for their meat, were heavily harvested for over 300 years. After the initial settlement of the island and up to the early years of the twentieth century, large numbers of females were taken on the nesting beaches (the males do not come ashore). They were 'turned turtle', which rendered them completely helpless and later collected at leisure, transported round the coast to Georgetown by sea and kept until needed in stone-built turtle ponds just outside the settlement. Sluices were built into the ponds which allowed the ingress of sea water while simultaneously preventing the turtles from escaping. Throughout its tropical range this turtle is still valued for food more than any other and ruthlessly harvested in many places, but the Ascension population is probably now the best protected in the world.

Green turtles are the most familiar and largest of the hard-shelled turtles, adults on Ascension measuring up to 3 feet (1 m) in length and weighing as much as 300 lbs (136 kg) During the breeding season the females come ashore at night, eight or nine times, excavating a hollow each time in which to lay clutches of 100 or more eggs. The number of eggs laid and thus the number of potential young is huge, but the mortality of hatchlings is enormous and it is only a tiny minority that survive to grow during the following years at sea, and, in time, to return as adults. Once away from the nesting island, the turtles are great ocean travellers and, with infallible navigational ability, cover enormous distances between visits to the breeding beach, intervals of which are anywhere between three and five years. Adults from Ascension regularly make the 1,200 mile (1,900 km) crossing of the South Atlantic to feed on sea grass meadows in the shallow waters off the coast of Brazil where they are well protected. Females make the return journey irregularly, but males do the crossing every year to ensure their availability at the tiny dot of land in the breeding season.

In addition to the important population of green turtles, there is also a small but significant population of hawksbill turtles which forage around the island and presumably use it as a staging post between feeding areas in Brazil and breeding sites on the West African coast. It is a humbling thought that both these ancient reptiles, denizens of the open ocean, have been riding the waters of the South Atlantic for aeons, long before Ascension first appeared above the waves five or six million years ago.

Fernando de Noronha

This oceanic island is as near to tropical perfection as one can imagine. Fernando de Noronha is an island of stunning beauty, its radiant coastline and turquoise seas punctuated with beaches of heart-aching perfection, some of them crouching below golden-yellow cliffs several hundred feet high with their feet studded with forest trees reaching high for the sky. Noronha is not a large island, being only 6 miles (10 km) in length and no more than 2 miles (3.2 km) at its widest. It lies 150 miles (240 km) out in the southern Atlantic from the Brazilian mainland, north-east of the city of Natal, almost 4° south of the equator.

At one time the island was completely forested, but much of it was felled in the nineteenth century, when the island was used as a prison, to prevent the inmates from hiding or building rafts in attempts to escape. Half the surface is still covered in forest although most of this is now secondary growth smothered in creepers, but it is nonetheless liberally scattered with the brilliant marigold flowers of mulungu trees, as startlingly bright as 'flowers of the forest' in Amazonia. However, the physical feature that towers above the island and appears in every picture is a remarkable volcanic plug, of phallic proportions, which dominates the skyline from every viewpoint. The Morro do Pico, on the central north coast, rises 1,050 ft (321 m) high above the perfect beaches which lie below it and dwarfs the surrounding forest vegetation. It is the unmistakable and unique totem that proclaims the island's identity and simultaneously defines its origin.

Twenty tiny satellite islets lie around the shores of the main island which thereby qualify the group as an archipelago. The seas around the islands are exceptionally rich in marine life, a fact that was reflected in the Brazilian government's declaration of the area as a major marine national park in 1988. The park extends to 45 square miles (117 km²), and also includes 70 per cent of the main island, the excluded areas being principally those where human occupation and development have taken over. The international importance of the archipelago was further recognized in 2001 when the whole of the marine national park was declared a World Heritage Site by UNESCO.

The wealth of underwater life around the island encourages spectacular, but closely controlled, diving and snorkelling in the bright clear waters. Unpolluted by sediments from any of Brazil's large rivers, the clarity of visibility down to depths of 100 ft (30 m) is a powerful visitor attraction. The waters support large populations of turtles, innumerable fish species, rays, skates, many sharks, as well as coral reefs, and ship wrecks. Tightly controlled tourism is the lifeblood of Noronha's modest economy, and there is an unmistakable feeling that, throughout all aspects of life on the island, everything is underscored with a consciousness of the need to nurture all the slender wildlife and other resources which the island possesses. There is little drinking water here; most has to be brought in. Similarly most other food supplies and all material goods are imported, along with oil which provides the power and lighting. Underlying it all is a universal appreciation of—together with a genuine fondness for—the bounty of wildlife that nature has bequeathed the islands and the care that is needed to protect it. For a start, there is a limit of 420 visitors on the island at any one time, all of whom pay a fairly hefty environmental tax on entry, which escalates with the length of stay.

In addition to the richness of the sea life, the island supports extensive wildlife in the form of endemic plants, several terrestrial birds, and one or two terrestrial animal species. Seabirds abound around all the coasts, for this is the largest concentration of breeding seabirds in the western South Atlantic. Marauding frigatebirds hang menacingly over the little harbour day by day, seeking spilled morsels or the chance for thievery, while others maintain piratical patrols offshore around the numerous booby colonies. Brown boobies, sleek and dark, abandon their normal deep-water diving, to plunge-dive recklessly in the shallowest water along the tide-line, imperiously ignoring any humans using the same space. The innumerable black noddies often specialize in the bravado of cruising underneath the high arch of towering breakers, leaving it to the last death-defying moment to escape the pounding surf. White terns (fairy terns) and red-billed tropic birds cruise the sea-boards in the cliff-lined areas and red-footed boobies adorn their urban tree-top tenements—'gannets' breeding in tree-tops seem so ridiculous.

Noronha boasts two endemic land birds as well as the obligatory extinct (but so far undescribed) rail. The Noronha elaenia, which has a maximum population of no more than 750 individuals, is the most easterly representative of the huge Neotropical family of tyrant fly-catchers. Like many of the others in the family, this elaenia is a plain bird. The Noronha vireo is a similarly unspectacular bird and both species occur fairly commonly in the forested areas, scrubland, and in gardens. A third terrestrial species is the eared dove which is a very confiding bird, found in almost all habitats from beaches and forests to built up areas, gardens, and roadsides. On the wing the little doves appear stub-tailed and, with rapid wing beats, often look for all the world like small parakeets in flight.

Being an oceanic island which has never had a connection to a land mass, Noronha has no native mammals. Inevitably, sometime in the past, rats became established following human arrival, and are still regarded as pests. One splendidly unsuccessful attempt to control the rats was the introduction of Argentine black and white tegu lizards in the 1950s. These are large animals, some male individuals reaching as much as 4 feet (1.2 m) and in appearance are strongly reminiscent of the Asian monitor lizards, although in no way related; they represent a classic example of convergent evolution. The lizards are carnivores and seemed to be an ideal predator to reduce the rats. Unfortunately what was overlooked was the fact that the lizards are strictly diurnal and the rats equally determinedly nocturnal so the two never met! The rats continue unabated and the lizards are now regarded as pests themselves as they willingly feed on birds' eggs when possible. On much more modest scale, the indigenous Noronha skink—up to 8 inches (20 cm) long including the tail—is an attractive creature which is extremely numerous and frequently appears around human activity, in the expectation of finding food scraps.

Apart from the rats, the only other wild mammal on the island is the rock cavy, a strange member of the guinea pig family, vaguely resembling a rock hyrax, to which it is distantly related. It seems wrongly proportioned with its head appearing to be too large for its long-legged body. However, the little vegetarians are frequently to be found in quiet areas, particularly around rocky ground, enjoying the sun on some of the coastal cliff tops. They were introduced to the island by the military in the 1960s for sport shooting but they certainly would not offer much sport now as you can easily walk up to them wherever they are encountered.

However the emblematic species on Fernando de Noronha are the abundant turtles and the great gathering of spinner dolphins. It is these two marine animals—notwithstanding the claim that the island has the best beaches in Brazil—that draw visitors from the mainland throughout the year.

Turtles occur around the coasts at all times of year and can often be seen easily from vantage points on cliff tops, cruising and feeding in the pellucid waters far below. Snorkellers see them regularly. The green turtle is the only species that breeds here, the females coming ashore to lay eggs between December and June on the protected Praia do Leăo beach in the south-west of the island. The females return to their natal beaches every three years or so. They come ashore at night and excavate nesting hollows in the sand above the tidal reach. Hawksbill turtles are also relatively numerous in the inshore waters, although they do not breed here and the population comprises exclusively juvenile animals. Unlike the green turtles which feed on sea grasses and algae, the hawksbills feed particularly on sponges, which is interesting as many sponges are toxic to most potential grazers and also contain many glass-like spicules. The hawksbills are one of the few creatures that are immune to the sponges' toxins but the means by which they cope with the mass of spicules remains a mystery.

In recent years Brazil has had an exemplary record in the protection of turtles (and various other marine life) along its long coastline. TAMAR,[1] its landmark marine conservation programme, celebrated its thirtieth anniversary in 2010. The TAMAR project on Noronha is one of twenty-two along the Brazilian coast. Every known turtle beach and feeding area in Brazil is covered by one of these sites, with the exception of the long, remote coastline north of Natal towards the mouth of the Amazon, where it is accepted that there may well be sites as yet undiscovered. In this way all turtles in Brazilian inshore waters, whether breeding or feeding, are heavily protected. Twelve of the TAMAR sites have major information centres and the one on Noronha is a must-see for all visitors to the island. It has helpful staff, first rate displays, user-friendly games and life-sized models of all turtle species—including those that are only of rare occurrence—which encourage children to climb on them, hide inside 'turtle eggs', or engage with swarms of baby turtles on a sandy 'beach'. Education is the foundation of much of TAMAR's work. All the children at the school on Noronha are taught about turtles. In the 1990s many of the children were smartly uniformed volunteer turtle guides, helping with the protection, catching, weighing, and measuring of turtles and talking to visitors. Surprisingly, because the work appeared to be too close to employment, this popular initiative fell foul of Brazilian employment legislation and was terminated. Nonetheless, an alternative initiative is already under way. At the age of 13, each year-group, as it passes through the school, is given intensive teaching about the turtles by TAMAR staff and later, at the age of 16 or 17, they undertake a two-week course out of which four new guides are selected to form a new generation of 'Tamarzinhos' who will again participate in the fieldwork of TAMAR and interact with the public. The project's visionary education initiative is not restricted to children. Every evening of the year lectures and videos on turtles, dolphins, sharks, and other relevant subjects are held in the fine open-sided TAMAR lecture theatre. In my experience attendance at the lectures is remarkable, with audiences of adults and children—both residents and visitors—in the late evening frequently over spilling to lean in through the windows or sit on the steps and in the aisles.

Projeto TAMAR has run a long-term mark–recapture study of green turtles and hawksbills caught in the waters surrounding the island as part of a long-term research programme. This has helped to throw more light on the long-distance oceanic movements of both species. Sometimes the animals, caught by snorkelling TAMAR swimmers, are brought ashore to the beach where they are weighed, measured, and marked with metal tags with the public, who need little encouragement, crowding round enthusiastically to watch the process. At other times the turtles are processed under water which, the workers say, is actually easier than bringing them ashore. It is known (p. 76) that the Brazilian coast and shallow waters around islands like Noronha are an important feeding ground for green turtles from as far away as Ascension Island and that interchange is confirmed by tagging. What has also been discovered, however, is that the juvenile hawksbills make an even longer trans-oceanic migration. A juvenile tagged on Noronha in November 1998 was recaptured—when it was then recorded as an adult female—off the coast of Gabon, West Africa, in January 2005. This is a straight-line distance of at least 3,000 miles (4,800 km), the longest known journey for an individual of this species. This recovery has led to the suspicion that there may be important breeding sites, as yet undiscovered, for the rare hawksbills, somewhere in the Gulf of Guinea. Worryingly, many of the fishing communities in those areas of coastline are traditional turtle hunters.

The TAMAR work on Noronha does not stop with turtles, for it has initiated a pioneer programme of reafforestation with native trees. Until recently there were only seventeen known native tree species on the island. After intensive searches another three species have been found and all are now being cultivated in a nursery and replanted on experimental plots.

In addition to the turtle research, there is a second equally important programme under way. On the north-west side of the island is the Baia dos Golfinhos. This is the sanctuary bay, where all human access is excluded, for the largest known concentration of spinner dolphins in the world. Here, the daily routine of the dolphins' lives is played out. At dawn, as the sun rises, individual groups of dolphins enter the bay from the open ocean, and spend the day there resting, mating, protecting their young, and guarding against shark attacks. Then, later in the afternoon, perhaps around 3 pm, sometimes earlier, the groups purposefully move out again to spend the night feeding at sea. On some days the total of dolphins in the combined groups will number several hundred: at other times there may be as many as a thousand present. Much depends on the weather and sea conditions and the success of the feeding.

Projeto Golfinho Rotador is a research programme that has been studying these dolphins daily since 1990. From a vantage point on the cliffs overlooking the bay it is an inspiring sight to look down into the clear waters and watch the dolphins far below. The outlook from the viewpoint is an eye-watering wildlife spectacle. A semi-circle of cliffs frames two sides of the picturesque bay which in parts has many mature forest trees, particularly the mulungu trees blazing with the fiery colour of their flowers. All round the bay the trees are adorned with a multitude of white forms of red-footed boobies. The branches too are lined with legions of sooty-dark black noddies with smart white caps. Over the sea and against an azure sky, pairs of bride-white terns fly in the perfect synchrony of courtship flight, the bright sun lighting up their translucent wings. Fondly the islanders call them 'little brides'. Pairs of sombre-plumaged noddies too, cruise back and forth—'little widows' to the locals. Add to these all the passing boobies and frigatebirds and occasional red-billed tropic birds with their long white trains and the views of the dolphins are seen through a diaphanous veil of birds. Despite this constant avian activity, there is total peace; there is no sound of any birds calling to disturb the tranquillity of one of the world's tropical idylls.

The groups of dolphins are of varying size and much of the time many of the members are engaged in frenzied mating activities while others lie peacefully, resting on the surface. Spinners, the third most numerous dolphin in the world, are the only species that can rise vertically from the water and spin impressively on their long axis; skilled males will spin up to a maximum of seven times before landing. They also perform horizontal spins lower over the water and indulge in frequent tail slapping and rolling. These energetic activities help to make dolphin watching one of the most popular activities for visitors to Noronha. The ongoing research has thrown important new light on the purpose of these aerial gymnastics. Traditionally regarded as pure exuberance, the research now suggests that, rather than simply being imperatives of display, the eye-catching gyrations are important elements of communication both within and between groups. It is believed that it is not the aerial activity itself that is of primary importance, but the pattern of turbulence and bubbles created on landing that is used, together with vocalizations, to send messages from group to group or communicate intentions within the group. The composition of the groups is remarkably fluid and may change at will from day to day, but the fundamental unit within the population still remains the 'family', each under the

FIGURE 6 Spinner dolphin.
© David Fleetham/Bluegreenpictures.com

leadership of an older female, the matriarch. Across the other side of the world, on the west coast of the island of Hawaii is a bay with very similar configuration and dimensions to Baia Golfinhos. Here, in Kealakekua Bay, there is another large regular gathering of spinner dolphins, behaving in much the same way as on Noronha. The two research groups maintain contact and, unsurprisingly, are arriving at similar conclusions about the meaning of the behaviour of these engaging dolphins.

The dolphins in Baia Golfinhos make spectacular viewing but there is also a smaller gathering at the eastern end of the island close to an area where there is an impressive tide race round the headlands and islets. From here dolphins invariably come out to accompany boats, giving wonderful close-up views. The slow-moving boats give the animals no chance for bow-riding but research suggests that these groups of dolphins are entirely comprised of males and that they are accompanying the boats in the belief that they are helping to divert attention away from the females and young. Groups of males such as these can also offer effective—if not infallible—defence against shark attack. One strange and unexplained fact is that the only time the dolphins abandon their sanctuary in Baia Golfinhos for the day is on the odd occasions when an innocuous humpback whale enters the bay.

So, this is an island whose inhabitants put enormous store on the protection of their wildlife and for whom the education of their children and visitors about wildlife and other resources is a pre-requisite. Apart from a continuing assault on rats and wariness about the tegu lizards, there is only one other species that has caused concern. The attractive cattle egret, white plumaged with stooped gait, is arguably the most successful recent colonist of the bird world. It was originally native to southern Iberia, tropical and sub-tropical Africa, and sub-tropical Asia. From the end of the nineteenth century, with no human assistance, it began an enormous global expansion of its range, first into southern Africa and then steadily throughout most other parts of the tropical and temperate world, even reaching the remote islands of the sub-Antarctic. Today it is a familiar bird in all these places. After its arrival on Fernando de Noronha its numbers rapidly increased and it is now a numerous bird throughout all the open areas of the island. In places, such as short grass areas, roadsides, sports fields, and gardens, the numbers can be impressive, up to 100 at a time in favoured sites. It is difficult to imagine this successful and innocuous bird being unwelcome anywhere in the world. However, here it fell foul of the airport authorities amid concern about the risks of bird strike. Certainly the mown grass areas of the airport are like a magnet to the egrets. As a result several hundred birds are claimed to have been eliminated in the past few years but there are still several hundreds of these attractive white herons remaining to go on causing both pleasure to some and concern to others.

Mingulay

S ome islands evoke a magic simply through the sound of their name. Mingulay is one such. Ever since my youth, this distant island, with a captivating name, had a magnetic attraction for me. I marvelled at the thought that a remarkable human population had survived there, at the southern extremity of the Outer Hebrides—the 'Long Island'—and I yearned to see the swarming multitudes of seabirds which electrify the island in spring and summer. The lilting music of the name itself—'Mingulay'—is a siren call. All those who recognize the island's name almost certainly do so by virtue of the lyrical 'Mingulay Boat Song'. Ironically, the song has no association whatsoever with the island, save in name, but was written as a sea shanty in the 1930s by Hugh Robertson who, like many others, was seduced by the name although he never actually got there. Is Mingulay, an island which is an integral part of the British Isles, truly remote in the context of this book? Of course it depends how we, individually, interpret the word 'remote'. Certainly for the vast majority of people in Britain it is unknown, far-flung, and unreachable—and to most, not even recognized by name. Even for those who journey around our islands, it is rarely visited and is well off all the normal routes, at the tail end of the exquisitely beautiful chain of Hebridean islands.

For over fifty years I harboured the urge to visit this alluring isle. Many years ago I had an exuberant and ambitious correspondence with a wonderful old man, Archie MacDonald of Barra, then one of the co-owners of the island, who offered to take me there, but it came to naught and he died many years ago. The island, together with its neighbours Pabbay and Berneray, is now under the ownership and management of the National Trust for Scotland. Given perfect summer weather there is no more idyllic place on earth than the beautiful white sand beaches, machair lands pulsing with life, and the turquoise seas of the Outer Hebrides. The 9th of June 2006 was such a day. Donald Macleod runs his boat, *Boy James*, out of Barra during the summer for fishing trips or to take occasional interested visitors as far as Mingulay. As *Boy James* pulled out of the picturesque harbour and swung past Kisimul Castle, the softest of easterly breezes rippled the sea surface and the boat danced across the wavelets as we headed south. Even by early morning the sun, burning down from an azure sky, gave promise of an almost tropical day.

We sailed past the gleaming beaches and rocky headlands of Vatersay to starboard, struck west through Caolas Sandraigh (the Sandray Channel), and continued to skim south again, now leaving the inviting islands of Sandray and Pabbay to port and in no time were off the stacks at the north end of Mingulay. Our skipper, wind-tanned and sturdy, but typifying gentle Hebridean softness, was determined to exact as much as possible from a visit that, for me, had been so long in the waiting. He had promised that he would take *Boy James* close in along the wild west coast of the island before circling round the southern end to the landing beach on the east side.

The west side of Mingulay can only be properly appreciated from the sea. On any scale, its cliffs are stupendous and awesome. They are the third highest in Britain, (how few people have heard of them?) topped only by those at St John's Head in the Orkneys and beetling precipices of distant St Kilda. No morning light yet shone on their ink-black walls, save one piercing laser-shaft of sunlight which funnelled through a gully and rested in a bright pool on the silky sea. Almost within touching distance, the cliffs tower up and up and up, pock-marked here and there by caves, some at sea-level and some large enough to swallow a house, their recesses lost in an invisibility of Stygian blackness. The highest cliff here, Bagh nah-Aoineig soars up to a giddying 751 ft (229 m). Beside one of its flanks a rope stretches from cliff top down to the sea. But this is no relic of the former bird fowling by the islanders but the means by which climbers abseil down to sea level, then to test themselves on the challenge of climbing back to the top. The thought itself is numbing.

However severe the massive cliffs, they are certainly not devoid of life. On the contrary, they are one of the great seabird cities in the west of Britain and teem with a host of sea fowl. The massive stacks of Lianamuil and Arnamuil together with the adjoining mainland cliffs have been famous for their seabird throngs for centuries. Razorbills and guillemots process in constant streams to and from their improbably narrow ledges and the fishing grounds out at sea. There are some 12,500 pairs of the former and 9,500 of the latter.[1] They move silently as do the fulmars—some 8,700 pairs—turning in lazy wind-filled circles across the cliff face. All is by no means silence however, for the cliffs perpetually resonate with the shrill cacophony of onomatopoeic calls of 1,800 or more pairs of kittiwakes. They busy themselves at the nests, cemented to the tiniest of toe-holds in their seabird tenements, wedged onto the sheer sides of the cliffs. Visiting a large seabird colony at any time is a memorable experience; to see birds in such numbers as this on precipices like these leaves a truly indelible imprint.

When cruising these western cliffs of Mingulay, there is one final surprise. On the approach to the huge stack of Arnamuil, barely separated from the main island, the Ordnance Survey map identifies a 'natural arch', as it does for many similar features around our coasts. However, this is no ordinary arch and is not comparable to any other I have seen (save the remarkable Pearly Gates on Bear Island in the Barents Sea). Less an arch and more a marine tunnel, it is narrow, long, deep, and eerily threatening as an apparent cul-de-sac with no evident way out. However, at the far end and hidden from the approach, there is an abrupt right angle turn to the open sea. It is an exhilarating white-knuckle ride when sea conditions occasionally allow and a traverse is possible. Rising and falling on the swell, the boat eases into the tunnel, the walls literally within touching distance on either side and, swinging sharply to the safety of the open sea, shoots out at the far end. At any time it is a wonderful example of seamanship grafted onto a lifetime at sea on trawlers and other fishing boats in these challenging waters.

Mingulay has two faces and the eastern side is completely different from the steepling cliffs of the west. On a glorious summer day the broad crescent of white sand of Mingulay Bay shimmers above a silky sea and is backed by low dunes with the shadow of former in-bye land beyond. It is an image of Hebridean perfection. The usual landing is on the rocks at the north end of the beach from where one works one's way gently across the sandy slope. All is utterly still on days of perfect weather that, contrary to belief, are not unusual in these western isles in May and June. On low cliffs at the side of the bay pairs of fulmars sit, half hidden in hanging rock gardens of sea pinks, scurvy grass, and cascades of fescues. Pass nearby the fulmars and there are guttural complaints and anxious head swinging. Across the dunes the silence is disturbed by the sudden manic calling of an oystercatcher objecting to the intrusion and the softer, forlorn fluting of a ringed plover. Golden eagles breed here and with any luck can usually be seen. In the warm air one of the birds rises gloriously on thermals above MacPhee's Hill, head gleaming gold in the bright sun as it wheels in lazy circles in a master class of aerial control. It sinks behind the hill again and then rises high once more to sail majestically over its realm. The pair has a nest below the crest of the great chasm at Bagh nah-Aionoig, the Eagle's Cliff. For how many centuries have eagles bred on this mighty cliff?

On an eminence above the dunes behind the village, stand the forlorn gable ends of the little island chapel of St Columba.[2] Together with the schoolroom on the far side of the shallow valley, it is the most prominent remaining totem of the resolute community of islanders who won a meagre living from the harsh conditions of land, sea, and cliffs over the centuries. There is evidence of occupation on the island going back at least 5,000 years. A fortified promontory on the south-west cliffs probably dates back to the Iron Age. Celtic monks undoubtedly settled in these southern Hebridean islands from the sixth century and the Norsemen, 200 years or more later, have left their legacy in place names. Hecla ('hooded shroud'), is pure Norse; the last element of Skipis*dale* derives directly from the Norse word 'dalr' for valley, while Arnamuil reflects the Scandinavian name for the sea eagle—*erne*—and *mul* for headland. Even the word Mingulay itself is from the same era—*Mikil-ay*, the 'big island'.

In more recent centuries the 'modern' settlement was the village site at the base of the dunes behind the bay. The shadow of a head dyke built of stone and turf winds across the hill above the village, separating the former in-bye land from the rough open hill beyond. Downhill from the church the sad remnants of the village houses crouch, roofless, almost guiltily among nettles and other encroaching vegetation, their low walls and doorways now tenanted only by noisy families of starlings, some wrens, and an occasional pair of wheatears. A line of yellow iris traces one of the hidden runnels through the village and buttercups and ragged robin enliven the grassland. The whole site evokes an almost aching sadness: a memorial to untold years of privation, mind-numbing labour, and a constant daily struggle against an unyielding environment that would eventually win.

The history of the community on Mingulay over the 200 years up to the time of final evacuation is well recorded. The sandy patches of land behind the village were fertile enough, enhanced regularly with seaweed, animal manure, and ashes. The cultivated strips (the outlines of the lazy beds are still evident) grew crops of potatoes, barley, rye, oats, and roots. I sat on the rocks by the stream below the chapel, in the abandoned graveyard with its handful of identifiable gravestones. I tried to picture the scene at this time of year a hundred years ago when the first of the summer hay was taken as winter fodder for the few cattle, sheep, and ponies which were kept. Cultivated crops from the lazy beds were nowhere near sufficient to maintain the community and there was a critical reliance on fishing in spring and summer and the annual harvesting of seabirds and their eggs. The women did much of the heavy work with the cultivated crops and the stock while the men took to the sea or the even more dangerous task of scaling the cliffs for the harvest of birds. In the late eighteenth century there was a small community of fifty souls living on the island which rose steadily to 150 a hundred years later.

Around the British Isles, St Kilda (Chapter 9) is the most famous example of a community that had a total dependence on seabirds, where the fearlessness and climbing skills of its cragsmen are legendary. Mingulay is less well-known in this respect, but no less remarkable for the parallel feats that had to be undertaken to ensure the necessary harvest of birds and their eggs. Martin Martin, writing in 1695 recorded that the islanders 'take great numbers of Sea-Fowles...and salt them in the Ashes of burnt Sea-ware in Cows Hides'. Unlike on St Kilda, any and all species were taken wherever they could be reached—guillemot, razorbill, puffin, Manx shearwater, fulmar, shag, and kittiwake—and most were salted down to provide food through the harsh months of autumn and winter. The sea stack of Lianamuil, for example, is terraced with long narrow ledges packed with ranks of guillemots and others and was a principal colony, regularly visited. The men would climb the seemingly impossible cliff from the sea, often taking the birds on the ledges by hand, one after another, as they stood protecting their egg or young. At other times they used horse hair nooses on poles, wrung the necks and threw the birds to the boat waiting below. On at least one occasion as many as 2,000 guillemots were taken on Lianamuil in one day. Manx shearwaters bred on the grassy summit of the stack and the young were greatly valued and served as important part-payment for rents.

Puffins too could be caught by noose or manually hauled out of their burrows. (They were exterminated on the grassy summit of Arnamuil to protect the grassland for sheep grazing.) Apart from the primary purpose of taking birds for food, the feathers they provided were an additional valuable commodity. Feathers had an immediate use in pillows and mattresses but were also important as payment for rent and as a trading item. Astonishingly both Lianamuil and Arnamuil were used for sheep grazing. Until 1871 when it collapsed, a rope bridge spanned the chasm across to Lianamuil. Before that one man had to climb the stack from sea level and then haul the sheep up on ropes.

On my first visit to the island I badly wanted to see Lianamuil from land. I left the village and walked through the old in-bye land, swathed here and there with glittering carpets of silverweed. Did the islanders grow the plant long ago, I wondered, to use the roots as food in the days before potatoes arrived in the islands, as they did elsewhere in the Hebrides? The day was still silent and windless and there was a shimmering heat. A solitary skylark lifted its music high into the blue sky and a snipe spiralled out from between my feet with a harsh 'scaap'. In a damp hollow the tall brown stems of last year's reeds had weathered the winter storms and stood, seemingly defiant, as skeleton echoes of seasons past. Beyond the head dyke the land rose to a saddle between the slopes of Carnan on the south and MacPhee's Hill to the north. The acid heathland was crisp beneath my feet, a matrix of purple moor grass, cross-leaved heath and heather, dotted with the yellow stars of tormentil, the pink of heath spotted orchids, the parasitic lousewort and, here and there, the vivid lime-green traps of butterwort leaves, already flecked with the minute bodies of captive insect victims.

The view from the top of the cliff across to Lianamuil is awesome. The stack is huge, not only high—it is 403 ft (123 m) from the sea to its summit—but also massively broad, not pinnacled like those at St Kilda. The chasm between the stack and the mainland cliff is giddying to look into, among the whirlwind of birds around its

FIGURE 7 Lianamuil.
© Ben Buxton.

flanks. It is salutary to think of men working the stack through sheer necessity, year by year over the centuries. The width of the gap at the top is a few metres, a lovers' leap type of gap. I marvelled at the thought that there had been a horse-hair rope bridge across the chasm. It seems more likely that it may have been a single horse-hair rope secured at the mainland end which allowed a man to swing across the chasm to the opposite cliff.

The understanding of the challenge posed by somewhere like Lianamuil puts into context the deeds that life demanded from these people: just how impossibly hard it must have been. However, by the end of the nineteenth century the size of the population (150 in 1881) had become unsustainable. In addition, the island had always been cursed by the fact that there was no secure harbour or even landing slip, with the consequent difficulties of access, loading and unloading, and launching fishing boats. The dunes too were encroaching on the village, epidemics occurred and the conditions on the island gradually became increasingly intolerable. This was a community which, in the twentieth century world, was living too close to the edge. The land had always yielded its resources reluctantly: the crops in the lazy beds, the birds on the death-defying cliffs, and fish gleaned from the sea in a boat with no easy landing. For centuries the community had survived, laboured, and adapted, but by 1905 the first families were forced to emigrate and move to other islands. In 1911 there were only eight families still living there and in 1912 the last native inhabitants left the island for good to find refuge on the larger nearby island of Vatersay. Mingulay was finally deserted.

I made my way back to the landing, full of thoughts and images of a forgotten past. I sat on the rocks in the warm sun and waited for the dingy to come in. It was early evening. A puffin flew out from a burrow a few feet from where I sat. Another followed from further away and they flew together in a wide circle over the sea. Suddenly there were more and more at sea and, in a brief moment of time, the air was full of puffins racing round in front of the low cliffs in a silent mass of whirring wings. It was an apt farewell to Mingulay, an island now abandoned to the birds, to winter storms and deserted crofts— a wild and very beautiful island with its rich wildlife and a sad treasury of human memories.

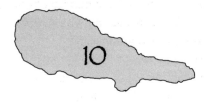

Pico in the Azores

As recently as the last quarter of the twentieth century people were intrigued by the knowledge that in the Azores, 1,000 miles (1,600 km) or more out into the Atlantic from the nearest point of mainland Europe, the island seamen still hunted the huge sperm whales in the age-old tradition of their ancestors. They pursued the leviathans in their fragile open boats, right up to the termination of whaling there in 1987. By this time, of course, commercial whaling everywhere else in the world had marched forward with technology, harpoon guns, metal hulled whale catchers, often accompanied by processing ships, but still the rugged Azoreans rowed out to hunt the 50 ton (51 tonne) sperm whales in cockleshell boats, as their forefathers had done 150 years before. Pico and Faial are the two principal whaling communities, where whaling has now turned full circle and boats still put out to sea to find the whales, but now taking eager tourists to see and photograph the great leviathans. Pico is the key place to visit to find out more about the whaling and it is possible to feel that one can almost 'touch' some of the history of this remarkable, residually primitive industry. It encourages an intimate relationship between man and some of the great creatures of the ocean.

However, there is a lot more to the wildlife of the Azores than their fame for dolphins, whales, and whaling. The islands are rich in birdlife, including many distinctive subspecies of birds we are more familiar with on the mainland of Europe: chaffinch, goldfinch, robin, and others. Another factor which lures European birdwatchers to these out-of-the-way islands, apart from interesting resident species and regular European migrants, is that the Azores are at a cross-roads in mid Atlantic, not much further from Newfoundland than they are from Gibraltar. Therefore they attract wind-drifted American species in spring and particularly autumn, as regularly as they are visited by wandering birds from Europe. Nearctic specialities such as American wigeon, ring-billed gull, killdeer, lesser yellow-legs, Tennessee warbler, and a host of others, are east coast migrants on the American continent, which, when blown off course across the sea, are fortunate if they make landfall on these islands in mid Atlantic. Pico and its neighbours now start to benefit from an increasing volume of birdwatchers from both sides of the Atlantic to add to other visitors who come to the islands for whale watching.

There are nine islands in the Azores archipelago, scattered over nearly 400 miles (650 km) of the ocean. These are volcanic islands, as is immediately evident when you land on any of them, for they are the summits of part of the massive undersea mountain range that is the Mid-Atlantic Ridge. Pico is the largest of the central group, the most recent and the most obviously volcanic of them all, dominated as it is by the towering Pico Mountain, 7,713 ft (2,351 m) high and as perfect a cone as Fujiyama or Etna. Life must indeed have been hard for the earliest settlers on Pico because, unlike the other islands, there is desperately little flat land and the hard-won cultivated areas are confined to the narrowest of marginal strips around the coastline. Especially at the western end of the island, both north and south of the little town of Madelena, the lava flows have left an all-encompassing landscape of black, razor-sharp lava. The labour that was involved in producing the innumerable, tiny, individual areas of cultivation in conditions such as these is mind numbing. But, volcanic soil, if it can be harnessed, is rich and fertile. Everywhere at this western end, lava rocks have been quarried, broken and worked to build thousands upon thousands of tiny walled enclosures, square ones for vines and round ones for figs, to offer protection from the sea wind and to help generate warmth. Pico was justly famous for its 'Pico Madeira' wine which was the staple of the economy until the vines were crippled by disease in 1820 and then finally devastated by *phylloxera* aphids in 1853. In human terms, that disaster caused untold hardship among the population on an island with desperately little alternative agricultural potential and it resulted in mass emigrations to Brazil and California. One hundred and fifty years later vast numbers of the abandoned enclosures have sunk ignominiously under a tide of encroaching vegetation, an unworthy end to extreme human labour. The most impressive remaining lava-walled vine enclosures are found south of Madelena. Here, you may stop on the roadside and look over an extensive stretch of country towards the sea, completely covered in a labyrinthine network of interlocked enclosures. It is a remarkable sight and, such is the historical importance of this area that UNESCO has designated it as a World Heritage Site, testament to the legacy of man's resourcefulness and determination in grimly challenging conditions. It is good news that in recent years the Pico wine industry has been rejuvenated, with new methods of cultivation, and once again produces an extremely palatable white verdelho wine.

The road along the north of the island passes the little airport and hugs the coast part of the way as it crosses a desert of bare black lava which tumbles into the sea over ferocious, low, jagged cliffs. Across the water is the long, whale-back outline of St Jorge and on a sunny day the sea between the two islands sparkles. Yellow-legged gulls, recently determined as a separate species having previously been regarded as a race of the herring gull, soar on the updraughts. Out at sea the flashing white undersides of Cory's shearwaters and occasional Manx shearwaters catches the light as they arc above the waves.

Away from the coast and heading inland, a cobweb of roads and tracks crosses the flanks of the mountain. Still the top of the mighty Pico Mountain remains unseen, guiltily hidden in a cloak of white cloud. People sometimes come for a week, especially to see it, but never even get a glimpse, shrouded as it is for much of the time under its thick canopy. The flanks of the mountain fall steeply to the sea, north, west, and south. The slopes below 900 ft (300 m) or so are a chequerboard of planted forest and open pasture. Higher up the European Union has financed the removal of much of the natural shrub cover, replacing it with poor quality cattle grasslands. It is only in the east where the long slope dips more gently to a distant headland, that one gets a real feel for the inland part of the island. Even at these altitudes low cloud and mist are commonplace at any time of year. Colourful hydrangeas grow wild all alongside the lanes, as they do throughout all the Azores. Birds are plentiful everywhere. A buzzard drifts out of the drizzling fog, swings in a lazy circle, and disappears just as quickly in the mist. Blackbirds and canaries are abundant, as are grey wagtails—one of the commonest birds in the Azores. Otherwise it is strangely wild and empty up here on the flanks of the mountain save for the omnipresent cattle grazing in the mists, often accompanied by small flocks of starlings. The total silence is lent a strangled feeling by the misty surrounds. On a fencepost by the road a snipe sits, hunched, and calls with a persistent 'chip-per, chip-per, chip-per', breaking the silence, teasingly unconcerned by any human presence.

Leaving the mountain tracks and returning westwards towards Madalena, the road descends bullet-straight for a few miles giving, under the cloud, magnificent views over the west end of the island and across the sea to its smaller neighbour, Faial. Madalena is a neat ordered town with patterned black and white buildings round the town square dominated by the church of Santa Maria Madalena. On the outskirts of the town one of the popular restaurants is above the cliffs, close by the sea. Here, after dark, one can sit at an outside table, lit by occasional street lights and listen to the wild night-time caterwauling of Cory's shearwaters as they home in to their cliff-hole nest in the dark. You can see them too, for they are easily visible as their white undersides flash as they pass through the glow of the street lights. Some of them land in the rough ground in fenced-off gardens alongside the cliff. How much would some birdwatchers give to have Cory's shearwaters nesting in their garden! In the day-time the same cliffs are peppered with the comings and goings of rock doves and the dazzling whiteness of little egrets catches the eye on the shoreline below.

The town crouches comfortably around its busy harbour. The ferry from here plies back and forth to Faial and this is the home port of an important tuna fishing fleet. It never was a whaling port; that distinction belonged to Lajes (although there were also whale boats and a factory at Sao Roque on the north coast). As boats move in and out of Madalena's harbour they pass by two huge rocks, rising like the pillars of a long-vanished giant gateway to the island. Tenanted by rock doves and yellow-legged gulls—and doubtless shearwaters—their popular names, Deitado ('Standing Up') and Em-pe ('Lying Down'), neatly describe their forms.

The collapse of the vineyards in the nineteenth century put an even greater imperative on whaling and fishing for those families that stayed behind when others emigrated to the New World. There were commercial whaling operations on several of the islands, Sao Miguel, Flores, but the most important ones were Faial and Pico. The great advantage of Pico was that the really deep water, in excess of 3,000 ft (1,000 m), where the sperm whales feed, lies within a mile or two of the shore whereas boats from other islands might have to go five times that distance to find similar depths. Azorean whaling was perilous, but it was also necessarily a highly refined and tightly organized operation. The first link in the chain were the 'vigias', spotters whose lifetime job was to man the observation posts dotted around the coast, searching the seas for the sign of a whale spouting and then sending the message to the men waiting by the boats in the harbour. The whale boats (*canoas*), 40 ft (12 m) long, alarmingly shallow and slender but obviously sturdy, were rowed with six men, including the harpooner, pulling the oars, and a helmsman. 'Baleia!'—when the word was given they were pulled out to sea by a tow boat until they were within half a mile of the whale, after which they cast off, stepped the mast, raised a sail and it was left to the rowers, with muffled oars, to approach the whale from behind in complete silence, as it surfaced. For the last hundred yards or so even muffled oars were deemed too noisy and the crew shipped oars and resorted to paddles until the boat was virtually atop the whale and the harpooner could brace himself against the thigh board and, with all his force, hurl the barbed harpoon deep to the whale's spine. This was a crucial moment for the whalers in their fragile craft as the leviathan thrashed in pain and frustration before plunging to the deep, spinning out 2,000 ft (600 m) of rope around the bollard on the boat. As the whale resurfaced and the line was drawn in, it raced across the surface, spinning the frail whale boat in its wake; in its time this furious surfing behind the whale was the fastest that man had ever travelled at sea! Eventually as the whale tired, the harpooner was ready again with razor sharp lances aimed at the vital organs. In time the exhausted animal gave up, rested in a pool of reddened seawater and was eventually towed back to the whale station.

Although the commercial justification for Azorean whaling was really finished by the 1960s it continued in the same hazardous way for another twenty years, not least because, on islands such as Pico and Faial, it was an integral and traditional part of the social and cultural fabric. It is what defined the men of Pico and Faial; it was their life and their being. For the past twenty years or more, the same families whose lives were formerly inextricably bound up with whales are again involved with the same animals albeit for a different purpose. These days, the whales are approached as before, but now simply with awe and cameras.

Lajes, half way along the south coast of Pico is the most famous centre for whale watching in the Azores. The road from Madalena passes by the historic vineyards, winds through colourful villages across the steep southern slopes of the mountain. In parts there are forest plantations on either side of the road, dark tunnels in an otherwise wide-open landscape. Along one of these there are Azorean noctule bats, regularly flying in broad daylight. They are unique in being the only known insect-eating bat in the world to hunt principally by day, while most bats are nocturnal to avoid predation by hawks and falcons.

Lajes itself is no more than a small village, dedicated now to a summer flow of north Europeans coming for the single purpose of going to sea with the whales. The entire village is little more than one street, a small industrial park and a harbour front. Even before the season for tourist boats begins, there is no escaping the sense that the entire village revolves around whales. A first rate Whalers' Museum has been developed in nineteenth-century boathouses on the harbour front and contains a wide range of artefacts and information about the old whaling customs. There are whale-watching billboards and boat advertisements everywhere. Whale and dolphin motifs decorate the shop fronts, the bars and even the Moby-Dick shaped food stand on the harbour front. Life on Pico has always been hard and communities such as this deserve the fortune that nature has unexpectedly handed to them. The inherited skills of generations are still harnessed in traditional ways and the men of Lajes maintain their partnership with the whales and the sea.

Pico is an island of rare beauty, a powerful atmosphere, and an exciting range of marine and terrestrial wildlife but, above all, it is imbued with the essence of a proud heritage, inextricably married to a dependence on the great whales.

Tristan da Cunha

It is all too easy to deceive ourselves with a rosy image of carefree, relaxed, and contented lives enjoyed by small communities on remote islands, isolated, as we imagine, from the pressures, conflicts, and stresses of life in the outside world. Actuality is frequently very different. So often the human story of these far-flung islands is one of unremitting hardship, declining resources, and isolation, and increasing disenchantment eventually leading to abandonment. It is a decline which is often fuelled by a realization of the life styles and facilities that are enjoyed in modern times by much of the rest of the world: around the British coasts alone, St Kilda (Chapter 13) and Mingulay (Chapter 9) are two classic examples of communities that were eventually forced to give up. Interestingly, the human history on Tristan is almost exactly the reverse. Here, uniquely, less than 200 years ago, a small band of Europeans took the positive step of colonizing the island and establishing a permanent settlement which flourishes to this day. Tristan da Cunha is famed as the most remote inhabited island on the planet, lying in the temperate waters of the mid Atlantic, 37° south of the equator, 1,750 miles (2,800 km) from South Africa and 2,000 miles (3,200 km) from the nearest point of South America.

Even in its relatively short relationship with man, Tristan has a singularly interesting history in the annals of human occupation on remote islands. It was first discovered by the Portuguese in 1509 and later used as a staging post by whalers and sealers in the eighteenth century. Even in those early days the whalers, on their way south to the sub-Antarctic, did not ignore the opportunity of exploiting the populations of fur seals and elephant seals that abounded there. In 1810 the island was annexed by a New England adventurer, Jonathan Lambert, who proclaimed that he had 'this day taken absolute possession of…Tristan da Cunha and (its satellites) Inaccessible and Nightingale islands'. However his bold pioneering tenure was short-lived and barely successful, as he and two companions were drowned at sea two years later, leaving a fourth member of the group on shore who was still there when the advance party of a British garrison arrived in 1816. The following year the main garrison was billeted there temporarily, to prevent any attempt by the French to use Tristan as a base from which to rescue Napoleon from St Helena, echoing the simultaneous establishment of the first settlement on Ascension Island (Chapter 7). Three members of the garrison eventually opted to remain on Tristan, under the leadership of William Glass, a Scotsman from Kelso. A few years later those three were joined by several other men. The problem of the absence of women on this remote isle was neatly solved by persuading a ship's captain to seek wives for them on his next voyage from St Helena. Despite the understandable reluctance of eligible candidates on St Helena to travel to an unknown island they had never heard of, to marry men of whom they had no knowledge, he eventually recruited five volunteers (and the two daughters of one of them), deposited them one morning on Tristan and abruptly left to avoid the possible embarrassment of having them rejected and being left with them on board. Over the years the resulting population has slowly risen to a present level just below 300. It is a telling fact that the family names on the island are still those of the nineteenth-century pioneers—Glass, Repetto, Green, Swain, Hagan, Rogers, and Lavarello.

Tristan da Cunha is the principal island of an archipelago which includes four other nearby isles, the largest of which are Nightingale and Inaccessible, but it also technically includes distant Gough Island, 110 miles (180 km) to the south-east: all of these satellite islands are uninhabited (bar a small meteorological station on Gough). The main island of Tristan is essentially one huge volcanic cone, rising from the abyssal depths of the ocean, to a height of 6,765 ft (2,062 m) above sea-level, cliff-girt for much of its 20 mile (32 km) coastline, interspersed only with occasional coastal strips and relatively few stony beaches. The one small settlement is Edinburgh-of-the-Seven-Seas (not named after the Scottish capital, but after an earlier duke of the same title who landed there in 1867). It occupies a site on the northern tip of the island, near the eastern end of the largest lowland plain. The tiny Calshot harbour—the only safe landing place for visitors to the island—was built in 1965 around a small reef immediately below the settlement. Lying far out in the broad Atlantic, unsheltered by any nearby lands, the tiny harbour regularly suffers the full force of the sea, is frequently in need of maintenance and remedial work but remains the crucial link with the rest of the world. Prior to the construction of the harbour, landing was normally made on one of two open beaches—Big Beach was the principal one—until it was lost under the lava flow of a 1961 volcanic eruption.

However, man's arrival inevitably signalled the beginning of the exploitation of the island's abundant natural resources, leading to the catastrophic decline of many of the bird species which nest there and much of the native vegetation on the island. Over the succeeding 200 years or so the island's environment and the enormous richness of its wildlife has been severely depleted. The natural vegetation on the scattered areas of lowland around the coast has been completely stripped. Indigenous tussock grass which characterized these salt-spray areas has been almost completely eliminated by grazing stock. Although the tussock was later replaced by introduced grasses, they too were subsequently just as severely overgrazed. The rising ground above the small lowland plains is dominated by spectacular tree fern *(Blechnum sp)* growing up to 13–16 ft (4–5 m) and interspersed with the 'island tree' (more widely known as island cape myrtle), although the tree has long been felled for fuel and is now absent from many areas, especially those nearest the settlement. Even in places sheltered from the persistent winds, the tree grows to a maximum of no more than 13 ft (4 m) and is more often low to the ground with a gnarled trunk and procumbent form. This Tristan subspecies has a disjunct distribution, being found only on the Tristan group, including Gough Island and—remarkably—3,000 miles (4,800 km) away at exactly the same latitude on Amsterdam Island and St Paul's Island in the southern Indian Ocean.

With its abundant natural resources, Tristan had evident attraction for prospective settlers. It possessed ample fresh water and provided a seemingly limitless supply of fresh food in the form of abundant birds and sea mammals. Its main disadvantages have always remained its extreme distance from the mainland and regular difficulties in landing. Landings on the rocky foreshore left after the 1961 volcano eruption, were hazardous on anything but the rarest of calm days and even now there are many occasions when a landing by small boat into Calshot harbour is not safe. The harbour entrance is necessarily narrow to provide as much shelter as possible.

On one breezy day at the end of October we lay offshore hoping for the sea conditions to improve. The sky above and around was alive with birds, exquisitely beautiful sooty albatrosses riding the wind with consummate ease and a grace and elegance that belies their size, and shearwaters, prions, and petrels lifting and swinging over the white-crested swell. Eventually our zodiac was in the water and we stood off a couple of hundred metres from shore, watching with interest the breakers crashing against the shoreline and the harbour defences and, as is so often the case, awaiting a signal from the harbour wall for the right moment. When a raised arm signal eventually came, it was 'full ahead' and we surfed in on the crest of a breaker and, to our relief, made it inside to the relative calm of the tiny concrete harbour.

Above the harbour and away from the sea's tumult, Edinburgh (the settlement, as the islanders call it) has a quiet, friendly, and comfortable feel. To the west of the settlement, in the shadow of the great mountain, an undulating lowland plain stretches four miles or so, passing on the way the long-established potato patches, each one surrounded by lava-block dry walls, where the planted ground is fertilized each year with the offal from the crayfish industry. Cray fishing is the economic driver of Tristan, the catch being prepared and frozen in the island canning factory and exported through South Africa as Tristan rock lobster. The importance of the crayfish to the community is reflected in its prominence on the island's coat of arms and flag; an unusual distinction for a crustacean. The crayfish is enormously numerous in the surrounding waters where it inhabits the rocky shelves around all the islands of the group, optimally at depths of 65–130 ft (20–40 m). This subspecies is endemic on Tristan and, puzzlingly, also on the Vema sea mount 1,050 miles (1,600 km) north-east of Tristan. The oldest method of catching the lobsters was by the simple expedient of lowering a weighted line to the desired depth on the submarine shelf and drawing it up again minutes later with lobsters clinging manfully to the bait. Even at the maximum catch of 158 metric tonnes (2010–2011 quota), this is one species for which annual levels of harvesting have so far proved wholly sustainable.

Beyond the potato patches and a further couple of miles along the flat land, a tortuous path leads up the side of the mountain. It is a severe climb, initially up the slippery side of one of the many deep gulches that radiate from the mountain and, higher up, straight through the vegetation on the slope of the volcano. It can indeed prove difficult and, without the help of fixed ropes as it gets even steeper towards the top of the climb, it is doubtful that many would make it. Eventually at around 1,970 ft (600 m), the path emerges onto the Base, the local name for the wide, deeply dissected plateau from the centre of which the cone itself reaches to the sky for another 5,000 ft (1,500 m). The Base, heavily vegetated with impenetrable tree ferns where one could get instantly lost, evokes haunting parallels to the undiscovered primevalism of Conan Doyle's *Lost World*. It is comprehensively a world apart from the overgrazed and managed lands far below. The aerial views from the Base across those lowlands and the distant settlement, out to sea and to the other islands are spectacular. Yellow-nosed albatrosses float by noiselessly, swinging effortlessly on the wind, barely turning to look at the human intruders and so close that on a calm day one can hear the sough of the wind through feathers. The incubating birds sit patiently on their nests, almost insultingly heedless of human approach. These are 'mollymawks' (or 'mollies'), to give them the colloquial name— the name by which they and their near relatives are universally known throughout the southern oceans. They sit motionless on their nests with heads upright; white-bodied and black-winged. Their dark eyes and shaded brows lend them a permanently frowning expression. The black bills are large and powerful, pink-tipped and with the upper surface of the bill striped a signature yellow to give the majestic bird its name.

The lack of escape-reaction that these birds display and total lack of fear, renders them extremely easy to take and that has been their fate on Tristan over the centuries. It was always one of the great food birds for the islanders and its population has thus been dramatically reduced, although the species is now protected and numbers are starting to slowly rise again. The even larger Tristan wandering albatross has been lost completely from the island, having been eaten out of existence in the past and is now represented by only a couple of pairs on nearby Inaccessible Island, with the remaining 1,000+ pairs of the entire world population living on distant Gough Island.

On Tristan many other species of breeding seabirds were traditionally killed for the pot: shearwaters, larger petrels and, among the land birds, the endemic Tristan moorhen ('island cock') which was deemed to be particularly tasty, with the result that it too was eaten out of existence in the nineteenth century. The same ultimate fate befell the spectacled petrel while populations of other nesting seabirds also collapsed dramatically following man's arrival and the inescapable introduction of black rats in 1882. Two further species traditionally featured importantly as food items for the islanders and were taken in huge numbers: northern rockhopper penguins and great shearwaters. The penguin nested on Tristan in one or two large colonies which were regularly plundered for eggs (and on a lesser scale for meat as fishing bait) and the species here is now greatly depleted which is disturbing since over 90 per cent of the world population breeds on the archipelago. On Gough Island the penguins have declined by more than 90 per cent from 2,000,000 pairs, mirroring a similar loss on Tristan from some 200,000 pairs 130 years ago. The story of the early years of man's colonization of Tristan da Cunha is yet another example of an island community which was very largely subsistent upon the bird populations with which they shared the island, although they were living a fallacy, believing blindly that the birds were so numerous as to be inexhaustible.

On nearby Nightingale Island over 2,000,000 pairs of great shearwaters nest under and amongst the dense cover of tussock grass. With the huge rockhopper colonies there, the island provided regular supplies of both eggs and meat. In years when clement conditions permitted, four visits could be made to Nightingale. In September new-laid penguin and albatross eggs were collected, together with adult great shearwaters, followed by a November visit, again to harvest the adult shearwaters. Trips for guano from the penguin colony for spreading on the potato patches sometimes took place from January to March and in April it was time for a harvest

PLATE 1
Brilliant colours of Arctic flowers (© author)

PLATE 2
Mykines: fleyging for puffins (© author)

PLATE 3 Faeroese slaughter of pilot whales (© Mads Emil Johansen/FaroeIslandsPhoto.com)

PLATE 4 Brown treesnake (© Daniel Vice)

PLATE 5
Guam rail with radio tracking aerial
(© Ginger Haddock/Fernbird Photography;
courtesy of Suzanne Medina)

PLATE 6 Kuna Indian on San Blas Islands (© author)

PLATE 7
Masked boobies nesting on Ascension Island (© Photographed by Ascension Island Government and RSPB)

PLATE 8 Monitoring turtles, Fernando de Noronha (© Armando Santos)

PLATE 9   Mingulay village (© Ben Buxton)

PLATE 10
Rediscovered annagramma fern, Ascension Island (© Ascension Island Government.
Photographed by Reinhard Mischke)

PLATE 11 Artificial eider nests, Vigur (© Tim Olson)

PLATE 12 Village street, St Kilda (© author)

PLATE 13   Bird Harvest, Nightingale Island (© Jim Flint)

PLATE 14   Polar bear (© author)

PLATE 15
Ornate day gecko (© Vikash Tataya)

PLATE 16 Beehive huts, Great Skellig (© author)

PLATE 17 Pink pigeon (© Vikash Tataya)

PLATE 18
Tuamotu kingfisher (© Kris Zawadka)

PLATE 19 Tuamotu sandpiper (© author)

PLATE 20 'Vini': Tuamotu Kingfisher (© Iolo Williams)

of young shearwaters to boil down for cooking oil; 20–25 young birds would produce a gallon of valuable domestic oil.

The daunting uncertainties of remote places such as Tristan are emphasized by the violent volcanic eruption that took place there in 1961. The eruption, from a vent alarmingly close to the settlement, resulted in the well-recorded evacuation of the island, most of the islanders, however, choosing to return from the UK two years later, once the eruption had quietened. The canning factory, the key to prosperity on the island, was one of the casualties, being buried under the lava flow. A replacement burned down in February 2008 and a third one is now in place.

When the islanders returned in 1963, after the eruption, a colleague, Jim Flint, went with them as schoolmaster. In September 1964 and 1965 he twice made the hazardous crossing to Nightingale with the islanders to harvest seabirds, in the canvas-covered longboats, and gives a graphic description of the experiences from an era now gone. He describes how, after days of waiting for the right weather and sea conditions to make it safe for the boats to set sail, there was great excitement throughout the island. The longboats, loaded with kit bags, drinking water, and crates for bringing home the eggs, were launched with great care to avoid any risk of tearing the thin canvas on the rocky beach. Once in the water strong arms quickly pulled the first boat beyond the ceaseless surf to await the others. Experience and memories of earlier near disasters resulted in the rule that no boat would ever go unaccompanied and that all four boats should stay as close together as possible the whole way over. Once on the open sea the mast was stepped, the rudder fitted, the sail raised and the flotilla was under way. In favourable conditions the 20 mile (32 km) crossing might take three hours but in calm weather it could be twice that and gave ample time for nerves to settle, anticipation to build, and reminiscences and anecdotes from the past to help pass the journey.

There is no beach on Nightingale and making land with the canvas boat was a delicate business for, with a swell running, it required men to leap onto the slippery landing rock, while the coxswain worked to maintain his station. All the cargo was manhandled ashore before the boat was finally pulled high up over the rocks and into the safety of a low sea-cliff and the tussock grass. Everywhere on Nightingale there were birds. Rockhoppers massed on and above the shoreline and far into the tussocks, their guttural calls drowning out the sound of the surf. The sky was full of albatrosses and thousands of shearwaters, the latter awaiting the failing light of day before making their return to land to find their nesting burrows. Where burrow space is exhausted the shearwaters lay their eggs in the open among hollows that fill every last yard of the ground. The September trips were exclusively to collect the fresh eggs of both penguins and albatrosses. Each man had a voluminous shirt tied tightly at the waist into which he gathered the eggs until he had his full 'bosom of eggs' and then returned them to the landing to be packed carefully in the large crates. Flint stressed that fighting one's way through the head-high tussock, collecting the eggs, was no pleasant stroll. Furthermore it was made infinitely worse by the evident wrath of the penguins that reacted to intrusion with powerful beaks and beating flippers. Stumbling or falling in the colony was not amusing and could be a very painful experience. Depending on the timing of egg laying, the collection might occupy as little as three days or so and a full catch resulted in what Flint calculated was a total of some 23,000 penguin eggs per annual trip and 1,500 of the thinner-shelled, more delicate albatross ones.

Whatever food was brought on the boats was supplemented on shore by an evening diet around camp fires of penguin eggs, boiled shearwater, and potatoes. Hundreds of the shearwaters were killed, skinned, and salted on the September trip for consumption later. Once the collecting was complete, it was a question of watching the sky and waiting for the weather to offer an opportunity to launch

FIGURE 8   Bosom of eggs.
© Jim Flint.

the boats and make a quick return to Tristan. The heavily laden boats sometimes made difficult passage. Flint recorded how on one return journey the wind was too strong, the sails were reefed and part of the journey was spent bailing as waves broke over the low gunwale of the open boat. However, despite anxieties over the years, nobody was ever lost on a Nightingale trip.

The huge populations of breeding birds on Nightingale are an echo of what the picture was like in the past on Tristan itself, and as it still is on the other uninhabited islands in the group—Inaccessible, Gough, and the two smaller islands, Stoltenhoff and Middle. The group as a whole is one of the most important seabird breeding stations in the world, with a total of somewhere in the order of 8,000,000 breeding seabird pairs of twenty-two species, many of them seriously endangered. Without question numbers were even higher than these 200 years ago. Including land birds, a total of twenty-nine species breed on the archipelago; of these, thirteen are endemic species or subspecies, including distinctive buntings on the four largest islands (although the Tristan one is now extinct), the Gough moorhen (introduced on Tristan in the 1950s and now prospering), and the Inaccessible Island rail, the smallest flightless bird in the world.

Distant Gough has its own problem: 99 per cent of the world populations of breeding Tristan albatrosses and Atlantic petrels occur on the island, together with the endemic Gough bunting. However in 2007 it was discovered that house mice, probably accidently introduced onto the island by whalers 150 years ago, had evolved in the way that island immigrants do, and have grown to some three times their original size. Moreover these 'super mice' have forsworn their normal vegetarian diet, turned carnivorous, and are feeding on the live albatross nestlings and petrel chicks at levels which, if left unchecked, could well lead to their extinction. The island is often quoted as the most important seabird breeding island in the world. The UK government has allocated funds to enable further research to be carried out on the island in the hope that the problem may eventually be overcome before it is too late.

On Tristan da Cunha itself, however, many aspects of human affairs and the wildlife have seen dramatic changes over recent decades. Tristan, of course, remains the most isolated human community in the world, although electronic connections to the outside world make isolation very different from earlier times. A regular service runs to the island from South Africa most months of the year. The six-day sea journey is undertaken by the fishing vessel MV *Edinburgh* six times a year and the research vessel SA *Agulhas* calls each September en route to and from Gough Island. Both ships carry passengers, mail, and supplies, while the exported crayfish continue to be the economic backbone of the island, despite recent market declines. Even tourism begins to feature more than might be imagined. A tourism officer has been appointed and a number of visitors briefly make it on to land each year from passing expedition ships as well as others who stay longer in recently provided holiday accommodation. Much has been achieved in arresting and reversing the environmental damage that occurred in past years. The island tree has been re-established in areas from which it was virtually exterminated and there are serious thoughts being given to attempting the major task of eradicating rats. Stocking levels too have been rationalized. Forty per cent of the island is prioritized for conservation and both Gough Island and Inaccessible are nature reserves. The whole island group is now designated as a World Heritage Site. On Nightingale young great shearwaters are collected and the eggs of rockhopper penguins are still harvested each year, but at levels that are regarded as acceptable to sustaining the populations. At the same time there are serious concerns about the northern rockhopper penguins, as it is known that throughout the archipelago (and more widely), their numbers have declined catastrophically for reasons that are currently not well understood, but may be related to climate change, the changes in the marine ecosystem, or over fishing.

Millions of pairs have disappeared; the vast population on Gough has fallen to a maximum of no more than 60,000 pairs while the total on the other islands combined has collapsed to between 40,000 and 50,000 pairs. The situation was further worsened by the disastrous wrecking of MS *Olivia* on Nightingale Island on 16 March 2011, spilling 1,500 tonnes of heavy crude, causing the oiling of thousands of the globally endangered northern rockhoppers together with other species and putting the crucial rock lobster industry at risk.

Tristan da Cunha rises more than a mile above the sea, the volcanic tip of a huge submarine mountain, in one of the most distant corners of the oceans. Winter snows and frequent cloud cover regularly obscure the summit of the mountain and rainfall is generous; the settlement receives up to 6 ft (2 m) in a year and a January temperature of 24°C is a very warm day. But the very name engenders an intoxicating fascination for a place that, for most people in the wide world, can never be more than a misty, unreachable island of dreams. For almost 200 years the island was carelessly raped of its vegetation, birds, and, at times, its seals and great cetaceans. At the end of the eighteenth century it was said that the island's coast swarmed with fur seals and elephant seals; in one year alone 5,660 skins were taken by one ship. By the 1960s all had gone. Any that had been left by the sealers were soon taken by the islanders to render down for oil for their lamps. In the 1830s 2,500 southern right whales were taken in only five years in the waters around the island. By the mid twentieth century the whales were scarcely to be seen and both seals were no more.

However, the productivity of the sea remains the heartbeat of the island's prosperity and the islanders, stockmen, and agriculturalists on land are also, by necessity, men of the sea. The past pillaging of the island's resources has been turned round with carefully structured agriculture, restoration of some of the natural vegetation, and legal protection of the recovering populations of birds. All land is communally owned and each household owns its own stock at controlled levels. In the way that all islands generate their own unique atmosphere, Tristan exudes a purposeful air reflecting a strong and vibrant community where, although life will always be tough, there is contentment, permanence, and the comforts of a rich communal life. Moreover there is a new pride in its unique wildlife which is being nurtured and maintained in a much better balance with the needs of the community.

Vigur

It is very rare to find an island where it can be claimed that man and wildlife live in a true symbiotic relationship, wherein both species gain material benefit from the presence of the other. There are endless examples around the world, as we see throughout the pages of this book, of islands where man has been the thoughtless and reckless exploiter of birds, mammals, reptiles, or plants. Less destructively, there are other examples (e.g. Mykines Chapter 4) where the regular harvesting of species has been carried out with a consciousness of the need to maintain populations. There are others too where, at very least, there is a realization that the resource is finite. To find examples where commensalism is genuine, long-established, and enduring, however, is indeed a rarity. Vigur is one shining instance of such an island, where there is a living example of harmonious co-existence.

Vigur is a small island which lies at the heart of the deep fjord coastline in the remote north-west peninsulas of Iceland, a half-hour sea crossing from the busy fishing port of Isafjörður, in the clear blue waters of the fjord of the same name. Iceland has many wilderness areas that are little visited and difficult of access, but these north-west peninsulas have always been recognized as the wildest and most remote parts of the country. They are mountainous and sparsely populated, but starkly beautiful. Across the waters to the east from Vigur is one of the country's smallest icecaps, Dranga-jökull, high up on the most northerly of the peninsulas' lonely mountain ranges. Down at sea level, Vigur is one of only three or four islands in the fjord, an emerald jewel, set in its deep blue waters. It is no more than 1 mile (1.6 km) long and only 400 yards (365 m) at its widest, low and flat with only one bluff at its northern end rising to 100 ft (30 m) or so. In the mid-summer days of perpetual daylight, the air here has a crisp and breathless quality and the low evening sun in the northern sky lights up the world with a radiant glow that is unique to these high latitudes. None of mainland Iceland is within the Arctic Circle, but Vigur is a mere 44 miles (71 km) south of that line and shares Arctic benefits such as long summer days and an enviable wealth of wildlife. The island is privately owned and farmed by the one family who live there year round, through the welcome months of daylight and warmth and the equivalent months of cold-ness and dark. The rich island sward supports a small number of Icelandic sheep and provides grazing for them and for the one or two milking cows. The island has been farmed since the dawn of settle-ment and by the present family since 1882. A working windmill was built on the island in 1840 and operated until 1917. It is the only one in Iceland, and is now owned by the National Museum as a national treasure, but maintained here, in its rightful place on Vigur.

The island story, however, is really all about eider ducks, for they abound here and Vigur is the third or fourth largest eider farm in Iceland and therefore in the world, with between 3,500 and 4,000 pairs of wholly wild eiders nesting every year. Æðey, farther down the same fjord, with c.5,000 nests is that much larger than Vigur. The fact that the ducks chose to nest in close proximity to humans is no accident, for where possible they instinctively select locations, such as farms or proximity to villages, for the benefit of the added protec-tion that a human presence gives. Throughout Iceland the duck's principal enemies are Arctic foxes and feral American mink, although on Vigur island there are no ground predators and the main threats to the eiders are from ravens and great black-backed gulls. One pair of ravens usually nests each year and is tolerated as it acts as a deterrent to other ravens. However, if they or the black-backs start to plunder the eiders unduly, they are removed. In return, of course, the harvest of eider down from the sitting ducks is gathered carefully from the nests and is a very valuable commodity indeed, representing the majority of the income for the island every year.

Common eiders normally return to their previous nesting area each year. The flamboyant males begin the courtship on the wintering grounds at sea, but there are many months to wait before the first eggs are laid on Vigur in the second half of May. The pairs form a monogamous relationship, but only for the one season, as it is not carried on across the years. Once a clutch of eggs is complete and the female starts to incubate she will seldom leave the nest for the entire duration of the 28-day incubation period. She uses the down from her breast to line the nest and to cover the first eggs that are laid before the clutch is complete; similarly she covers the clutch with the insulating down if she leaves the nest briefly during incubation. This is the crop that is then harvested once the eggs hatch and the ducklings leave the nest. During incubation the female is remarkably tame, closely approachable and not easily disturbed, but remains resolute, sitting on her eggs. By the end of incubation she may well have lost as much as 33 per cent of her previous weight and it is not unknown for a female to die of starvation during that time.

Young birds, breeding for the first time, also show a tendency to return to their natal area, with the result that there is a high degree of related breeding within a colony. This manifests itself in the fact that some females may readily lay eggs in neighbours' nests and, once the eggs hatch and the brood moves onto the sea, an individual female will frequently take charge of a large crèche of all-age ducklings from several adjacent nests. The tradition of farming common eiders and collecting the valuable down dates back to the arrival of Norse settlers in Iceland in the ninth century, bringing with them the tradition and the skills to go with it. Since then, eiders have always been fully protected in Iceland. Of particular interest on Vigur is the existence of artificial eider 'tenements', a series of ancient stone nest sites which were built at the head of the beach near the sheephouse, around 1800 AD and are still well used every year.

Eider down is the softest, lightest, and warmest down in the world and its insulating properties are superior to any other down. It is better than swan or goose down, for example, in that, after compression, it returns immediately to its original form. Take a bundle of down on Vigur when it has been cleaned, close your eyes and have it placed on your up-turned palms and you cannot tell which one it is on without looking; it is virtually weightless. Its special properties have been recognized and esteemed since time immemorial. Evidence of its use can be traced as far back as the Stone Age where

FIGURE 9 Eider duck.
© The author.

it has been found in archaeological sites in northern Scandinavia. It was accepted as tax revenue in the Middle Ages and featured in the accounts of the courts in various northern European countries. Because its thermal qualities are legendary, its value has always been high. Notwithstanding this, the world production of down declined in the twentieth century, partly through the introduction of much cheaper, though less effective, synthetic substitutes but also because of the extraordinarily high price it commanded. Before the 1917 revolution in Russia, the White Sea area was an important down-producing area, the Russian Orthodox Church produced the highest global volume of down annually (see Chapter 18). The total produced on Vigur is about 120 lbs (55 kg) per annum; it takes about 60 nests to yield 2 lbs (1 kg) of cleaned down. Whatever the cost, there will always be a demand for Vigur's eider down.

During the time of the year when they are in residence the eiders' well-being is the overriding concern. Even the sheep, which could accidentally tread on nests, were traditionally removed to spend the summer on the mainland, although that practice seldom applies now. They were taken over in a boat rowed by eight strong oars, in an example of what is almost reverse transhumance: island pastures in the winter and mainland meadows in the summer. Nowadays there are numerous day trippers who visit the island from Ísafjörður during the tourist season and provide a useful supplement to the island income, but on shore they are strictly guided or corralled to avoid any risk of damage or disturbance to the ducks. Having said that, every year there is invariably at least one nest by the narrow pathway from the little jetty, passed by successions of single file visitors, within touching distance of the incubating bird. The majority never even notice, while on the other hand it is probably the most photographed eider duck in the world!

Eiders are not alone on the island, however, for other seabirds also spend the summer here in great numbers. There is a huge colony of puffins, some 15,000 pairs and, as elsewhere in Iceland, large numbers are caught each year to provide readily-available food. Around 7,000 are usually taken for home consumption, 80 per cent of which are two and three year old birds rather than breeding individuals (first breeding is not usual before four years of age). The principal method of catching is by taking flying birds with a long pole and a net in much the same way as is described on Mykines (Chapter 4). Arctic terns nest here too, trans-global travellers flying vast distances each year to and from sub-Antarctic seas to nest here in the sanctuary of this predator-free island. Black guillemots are numerous: jet black in summer plumage with a fetching snow-white patch on their wings in summer and brilliant scarlet feet. Many pairs breed in hollows at the head of the shoreline by the farm, under buildings and among boulders; returning parent birds stand patiently nearby holding small fish for the young, until the coast is clear for their return to the nest. Redshanks, nervous and noisy, bob anxiously on small eminences or flit to and fro if visitors intrude on their territory.

So, above all, Vigur is an island of birds although eiders take precedence over all others. If puffins are the visible pulse of the island, the eiders will always be its living spirit—a friend of man and the source of a uniquely valuable product. Here the eider, St Cuthbert's duck, æðarfugl, is undisputed queen.

St Kilda

In the late morning of 29 August 1930 the fisheries protection vessel HMS *Harebell* raised anchor and slowly drew out of Village Bay on Hirta, St Kilda.

It is difficult to imagine a more emotional event than that which was symbolized by the departure of the *Harebell*. For, on board were the remaining 36 inhabitants of ancient community of St Kilda, leaving for the last time—and for some the first time also. The group, men, women, and children, watched from the stern of the ship as the island faded into the distance; some tears were shed, but all the silent thoughts were their own. They had eventually given up the battle and left behind only the legacy of a remarkable human history and a street of sad, abandoned dwellings. They were the descendants of a primitive community whose predecessors reach far back into the unrecorded mists of time. It was a community that had existed in isolation and, over the ages, had survived only because of the abundance of seabirds on St Kilda, on which their subsistence depended.

Distant St Kilda is the most remote archipelago off the western coasts of Europe. It comprises a group of four small, cliff-bound islands and a few formidable sea stacks. The largest island is Hirta

FIGURE 10 Fowler.
© National Trust for Scotland.

which possesses the only safe landing. For millennia, a human population eked out a life of hardship and frequent privation on St Kilda. At some stage in the forgotten past, Neolithic settlers here were replaced by pre-Christian Celtic immigrants, although whether this change happened through abandonment or by physical removal is not known. Later the Vikings too left their mark on Hirta in the form of strongly Nordic place names—Oiseval, Soay, Ruaival. Nevertheless, the eventual colonists were certainly of Celtic origin, as their names and their variant of the Gaelic language confirm.

The earliest written record we have of life on remote St Kilda dates from the mid sixteenth century. One hundred and fifty years later Martin Martin visited and left a startling account of life there. After him, many missionaries and other visitors, including clergy and schoolmasters have left their own graphic accounts of life on the island during the eighteenth and nineteenth centuries. In the twenty-first century it is difficult for us to comprehend the reality of the harshness of life and the complete isolation of these people, not only physically from the mainland, but from any knowledge of what comprised life in the outside world. The strain of survival and hard labour is etched on the faces of many of the women at an early age in photographs from the nineteenth century. As Steel (1975) has written, they knew nothing of 'normal life', didn't understand fruit or other food items, had not heard of farm animals (except their own sheep and cows), did not recognize trees, rabbits, or rats and were amazed by wheeled vehicles and urban dwellings. They lived, quite literally, in a world of their own and dependent on wildlife; their separation from the rest of civilization for hundreds of years was almost total. A strongly religious people, they were nonetheless noted for their deep superstitions, lack of initiative, stubborn resistance to change, and uncommendable lack of hygiene. There was never a question of their physical or mental strength and they were, to a man and woman, of unparalleled morality; crime on St Kilda simply didn't exist. Everybody, through necessity, worked for the good of everyone else and not for themselves.

The islands were owned for many generations, by MacLeod of MacLeod at Dunvegan on Skye, whom the islanders regarded and respected in almost deifying terms. Once a year they relied on the visit of the owner's factor to bring seed corn and salt. Through him, they paid for these essential goods and annual rent in the form of sacks of feathers and coarse, woven flannel. Other than that visit and the occasional fishing boat or other vessel that might happen to call, their only method of communication with the outside world was by the St Kilda mail boat. This comprised a hollowed block of driftwood carrying a message in a bottle attached to an inflated sheep's bladder mounted by a small red flag. Cast into the sea, it relied on the North Atlantic Drift to bear it toward the Hebridean coast and the hope that it would be found there and acted upon. This method of communication was used to plead for help in emergencies, when starvation or disease wracked the islanders, almost up to the time of the final evacuation.

Seabirds were life to the St Kildans. Without them there could be no existence so far out in the ocean from other land. However in addition to their total dependence on vast numbers of seabirds and their eggs, they also maintained a primitive agriculture. They had a few cows, but sheep were the most important animals, cross-bred Blackface/Cheviots kept on Hirta which provided both milk and meat and at the time of the evacuation well over 1,000 were transported off the island. A flock of ancient Soay sheep ran wild on the adjacent cliff-bound island of the same name. These primitive, small brown animals, descendants of those brought to St Kilda possibly by Neolithic farmers or by Norsemen around 900 AD were used by the islanders only for wool, not for meat. When the islanders left in 1930 107 Soays were individually caught and transferred from that island to Hirta and formed the basis of the wild population living there ever since. These sheep now represent a unique isolated genetic source.

The islanders' principal grain crop was barley, grown in small 'rips', similar to the runrig system on the mainland, with ownerships of individual rips rotated every three years. Some oats were also grown but with less success than with the barley. Although there is good soil on Hirta, fertilized by millennia of bird droppings, the climate is frequently dreadful and there were many years of crop failures.

However, it always was the ability to catch birds in a variety of ways, depending on the different species at the different seasons that enabled this unique community to survive. In that respect, despite the fact that some species were taken in large numbers every year, there was no evidence of a reduction in populations, not only because the birds were so numerous, but also given the sustainable nature of the harvesting. In any event, most seabirds are long lived and therefore capable of many years of breeding in which to produce successful offspring to replace the parents, even given the frequent removal of their eggs or young. In excess of 500,000 pairs of seabirds of fifteen species breed on the archipelago, including 20 per cent of the world's Atlantic gannets, (formerly) over a million pairs of puffins, 50,000+ pairs of fulmars, and in excess of 1 per cent of the European total of kittiwakes and guillemots. Four of these species, fulmar, puffin, guillemot, and gannet were the most important food items. Such was the critical importance of the ability to work on the cliffs that every male was automatically a fowler. They were born natural cragsmen and, from as early an age as three, would practise climbing on the walls and the houses; by 10 or 11 years of age they were on the cliffs and by 16 were fully fledged cragsmen. All the climbing was done barefoot and it is interesting that over the generations a marked evolution of their bodies took place. They developed almost prehensile, gripping toes, set wide apart and flexible, together with powerful ankles twice the thickness of normal. On the towering cliffs the fowlers were fearless, reckless and, in earlier days, certainly competitive. Inevitably there were occasional fatalities, but if a man went off the enormous cliffs there was no point in searching for him in the sea below, because survival was impossible. No other island community in the world has ever compared with the rigours and dangers of harvesting that was the St Kildans' daily task.

In the continual struggle to maintain their existence, the two most precious and irreplaceable possessions were the island boat and the horse-hair climbing ropes which were held in common. There was no nominated leader of the community but each day, except the Sabbath, all the men met in the street for the daily 'parliament', to determine the day's tasks. The principal rope was only taken out with the agreement of the 'parliament', and this was usually not before 11 August, when its strength would be tested the day before the traditional start of the climbing season. Bird harvesting in one form or another lasted some nine months of the year.

The gannet colony on St Kilda, is 4 miles (6.4 km) north-east of Hirta on the cliffs of Boreray, and the two adjacent stacks, Stac Lee and Stac an Armin. Together they form one of the largest gannet colonies in the world with over 60,000 pairs. To sail between them in the breeding season is to witness one of the great wildlife spectacles of the world. Not only are the cliffs papered white with the nesting adults, but the air is rent with their guttural callings and the sky above and all around is a breath-taking maelstrom of thousands upon thousands of birds. It is a sight which, once seen, can never be lost from memory. The birds return to the colony each year in late winter and early spring, which is when, on a calm day, the fowlers made their first visit. The two stacks rear out of the sea, smooth-sided, sea-sprayed, and seemingly unscalable. Despite this they were nonetheless climbed by two skilled cragsmen each year, who moreover made the ascent at night. The lead man had the crucial task of silently killing the sentinel bird (cf Mykines p. 48). Once that was achieved, the two men could quietly go round the ledges clubbing the sleeping birds or wringing their necks one by one and throwing them down into the sea below. Failure to silence the sentinel before he woke the colony would mean that the whole expedition failed.

Return visits were made in May for the collection of eggs from nearby Boreray and Stac an Armin, leaving Stac Lee to produce plenty of young gannets—'gugas'—for later harvesting. The men themselves regarded the descent of the slippery cliff with heavy boxes of eggs on their backs as the most treacherous of all their climbing activities. The numbers taken were often considerable and the weight of the boxes often as much as 150 lbs (68 kgs). In the more distant past up to 22,000 gugas were gathered in August but later it was usually less than a quarter of that number and, in the years immediately before evacuation, no more than 300–400. Apart from the good supplies of meat on these large birds, their other parts could be put to endless good use: 300 gannets would produce a feather mattress, 500 were needed to boil down for a valuable barrel of grease, their stomachs were useful as liquid carriers, and their hollow bones made for good pipe stems.

Important though the gannets were it was always said that 'without fulmars St Kilda would be no more'. In the sixteenth century St Kilda boasted the only colony of fulmars in British waters. From then on they became more and more important, not just for their meat, but also because of their economic value. Feathers and fulmar oil were highly valued on the mainland—oil for lamps, down for pillows, mattresses, etc., and grease for various uses. It was also regarded as a highly energizing food item, (we now know that it is rich in vitamins A and D), and produced valued ointments. However, the market for fulmar products dried up by the end of the nineteenth century with the development of paraffin lamps, alternative stuffing for pillows, and more modern medicines. Still, great numbers were taken for home consumption. A stable average of between 120 and 130 birds per person was taken each year, however much the human population fluctuated over the years. Collecting young birds on the cliffs began on 12 August and only lasted two weeks or so. Eggs were never taken, because unlike gannets and auks (guillemots and razorbills), fulmars do not relay. By mid August young birds were at their fattest prior to fledging. Harvesting the young fulmars was very dangerous work on the cliffs, whether it was approached from above or from a boat at the base of the cliff. On the massive cliff face of Conachair the pairs nest in great density and young fulmars are heavy birds, each one weighing up to 3 lbs (1.4 kg). Each fowler wrung the necks of some twenty birds at a time, before having them pulled up to the cliff top or throwing them down to a waiting boat and then moving on to the next ledge. This harvest was a strongly communal one with women and children all having responsibilities. Returning to the village from the cliffs, some women would carry the heavy loads on the long haul over rough ground to the village. After every catch the birds were divided equally between all the households in the community.

Puffins were always the most numerous birds on St Kilda with a population of over a million pairs (although numbers nowadays are much reduced). They were taken in their nesting burrows in early spring, often with the help of small dogs and this provided the first fresh meat since the previous autumn. They were also taken above ground, sometimes by stalking the birds carefully with a long rod

FIGURE 11 Fulmar.
© The author.

and lowering a running noose over their heads. A similar method used a series of horsehair nooses fixed to a firmly anchored rope or wooden block. Puffins are incurably curious and the first bird caught invariably attracts others to investigate; one woman reputedly caught 280 in one day by that method! It is perhaps interesting to note that the method used so successfully on the Faeroes (p. 45) never evolved on St Kilda. Puffins are very good eating, although not very large and therefore many were needed to keep the community fed. They were mainly eaten boiled or stewed, as with fulmar and guga, and often included with breakfast porridge to give flavour. The meat was usually tough and undercooked because the fires, fuelled with fulmar oil, were inefficient and seldom fully hot. The islanders ate eggs in huge quantities. There were 180 islanders in 1697 when Martin Martin was there and he estimated that something like 16,000 eggs were eaten each week during the season— c.80 per person.

The fourth species that was exploited in any numbers and regarded as particularly good eating was guillemot. The main colony is on a sea stack, Biorach, in the narrow channel between Hirta and Soay, off the north side of the main island. To take the harvest there involved another feat of climbing which was regarded as technically the most difficult of all. (If a man could not climb Biorach he was not fit to claim a wife.) The process here demanded guile and low cunning. The birds return to the ledges in March and at dusk the men, having climbed the cliffs, would chase them off and then imitate a guano-covered rock by crouching with their heads covered by a white sheet. When the birds returned in the gathering dawn and tried to land they could then be caught by hand, improbable though it seems, with luck at a rate of as many as 100 an hour.

Occasional curious tourists began to visit St Kilda in the nineteenth century and the islanders started to barter for a variety of items brought over with the ships. One of their most popular currencies were the blown eggs of a wide variety of the non-harvested species on the island. Oystercatchers, snipe, gulls, eider ducks, petrels, and others were collected, but the most valuable of all were the eggs of the tiny St Kilda Wren, a distinctive and slightly larger subspecies than the mainland bird.

One species of mammal which occurred on the island was so dependent on the human presence that it disappeared completely within a year of the final evacuation. The St Kilda house mouse (cf Mykines Chapter 4) evidently arrived in the distant past and evolved into a larger, distinct subspecies of the common mainland animal. The St Kilda field mouse, related to the familiar mainland wood mouse, is still plentiful and quickly took over the village area once the house mice died out. It is an extraordinary animal in that, over the evolution of time, it has increased in size until, at an adult weight of around 1.8 ounces (50 gm), it is approximately twice the size of mice on the mainland.

However, it was the fact that outsiders began to visit Hirta from the first quarter of that century that was a significant factor in the eventual collapse of the St Kilda civilization. The breakdown in isolation, a realization of better lives elsewhere, and recognition for the first time that needs could be supplied from afar, all slowly began to spell an inevitable end. Much of the earlier spirit and natural happiness of the islanders notably deteriorated. Furthermore, although the population was stable at around 80 souls from the middle of the nineteenth century up to 1920, it fell away after that date until there were barely enough men to work the cliffs. The eventual voluntary evacuation on 29 August 1930 was sadly inescapable.

St Kilda's civilization was unique. Its story and its final demise have become a part of living legend. Its people lived lives of almost unbelievable hardship, probably without realizing the real scale of this until the arrival of outsiders slowly began to open windows onto the outside world. Even then, and with the sporadic presence of resident clergy, nurse, and schoolmaster, they were stubbornly resistant to changes that would have eased their lives.

FIGURE 12   Harvest of birds.
© PA/PA Archive/Press Association Images.

In addition to the four bird species that the islanders relied on and the ones whose eggs were useful for bartering, St Kilda supports other important sea birds. At night the slopes of Carn Mor and Dun ring to the soft purring of storm petrels and the manic laughing call of Leach's petrels. Manx shearwaters too fill the night sky over those slopes with their weird cries. All make their visits to nesting burrows at night to avoid predation by the large gulls and the skuas. Nonetheless it is believed that the burgeoning population of great skuas on the island may now be responsible for killing large numbers of these small petrels each year.

Today St Kilda is a Mecca for naturalists and historians, not just from Britain but worldwide. Its bird populations and other wildlife flourish under the ownership of the National Trust for Scotland and it is protected under numerous designations, not least as a National Nature Reserve and its dual notification as World Heritage Site, first for its human artefacts and associations and secondly for its scientific importance. Each year a handful of visitors take the gamble of being able to make landing in Village Bay during the summer months. Visually, however, since the late 1950s, the island's powerful mystique is seriously compromised by the characterless intrusion of Ministry of Defence buildings above the landing and immediately below the old village street. It is an intrusion of almost sacrilegious proportions. A one-mile metalled road now winds up the 1,170 ft (357 m) mountain side to a military tracking site on the summit of Mullach Mor. The lower part of it traces the exhausting route over to Gleann Mor which, in summer, women and girls walked from the village morning and evening to take grass for the cattle on the way out and bring pails of milk on the journey back.

We do not know when St Kilda was first subject to human occupation, but archaeological research by the Royal Commission on the Ancient and Historical Monuments of Scotland on Hirta's satellite, Boreray, in 2011, has produced some astonishing findings. The grassy slopes of the island, which is no more than a third of a square mile (1 km²) in total are precipitously steep and its western face is a vertiginous cliff 1,260 ft (384 m) high. Consequently it has always been recognized as defying inhabitation, save for annual bird harvesting visits and shepherding from Hirta. Notwithstanding this, it has now been shown that there was indeed a resident agricultural community there with an extensive field system, crop terraces, and settlement mounds dating back to the Iron Age. We can all recognize the impossibly difficult life that defined the community on Hirta; the new discoveries on Boreray simply beggar belief.

South Georgia

On the 17 January 1775, during his second voyage of exploration, Captain Cook discovered this wind-swept island deep in the sub-Antarctic seas, mountainous and encrusted with ice and snow. Cook landed in a bay on the north side of the island which he appropriately named Possession Bay and, in a brief ceremony, claimed it for England 'in his majesty's name', thus giving the island its permanent name. Like visitors ever since, Cook was astonished by the uncountable throngs of sea mammals and birds that populated the open shores and the tumultuous surrounding seas. Once the word got out, within a decade British sealers descended on the island and, joined by American ships a few years later, entered on the devastating slaughter of elephant seals and Antarctic fur seals. Man had arrived and the inevitable despoliation of another pristine environment had begun. Within thirty-five years an astonishing total of 1,200,000 fur seals had been harvested for their skins and the species was on the brink of extinction on South Georgia. There would be worse to follow as the island's wildlife was to be pillaged further through the nineteenth and twentieth centuries.

South Georgia is a large island, some 100 miles (160 km) long and up to 24 miles (38 km) across at its widest. It is a hugely daunting, rugged but incredibly beautiful island, most of it under a permanent mantle of snow and ice, with a mountainous backbone rising to a dozen peaks of over 6,500 ft (2,000 m) and the highest just short of 10,000 ft (3,000 m). Its coastline is deeply indented and forbidding, with many inlets, precipitous headlands, and tide-water glaciers. It is bisected by the 54°30´ parallel but shares many of the characteristic features and conditions of Antarctic lands further south. Its icy waters are the most fertile and productive in the world, a fact amply reflected in the extraordinary wealth of its wildlife. It is, quite simply, for its size, the most remarkable place on earth for the enormous variety and incalculable numbers of its birds, mammals, and marine life. An unparalleled total of twenty-nine seabird species breed regularly on the island and its outliers (plus another four recorded only rarely), with an estimated grand total of over 35 million pairs.

The fur seals were only saved from extinction at the hands of the sealers by the fact that, by 1825, there were no longer enough to make killing worthwhile. By the 1870s a modest recovery in the population permitted limited numbers to be taken again until the last killing in 1907.[1] Since then, the fur seals have demonstrated the remarkable recuperative powers of wildlife, given freedom, with an impressive recovery to reach a population which is now probably in excess of 3 million animals. Elephant seals were also hunted mercilessly, although from the early 1900s it was at least carried out on a licensed and allegedly sustainable basis. Since hunting finally ended in 1965, they too have staged an impressive recovery, with somewhere around 400,000 individuals now inhabiting South Georgia. Seal hunting was necessarily a summer activity, for that is when seals are on shore both for breeding and for moulting. The sealers, therefore, did not overwinter on South Georgia which must have been a great relief to them because their makeshift accommodation in the summer months was rudimentary in the extreme and their existence even then, wretchedly miserable.

However, in 1904, a Norwegian whaling captain, C. A. Larsen landed in Cumberland Bay and set up the first Antarctic whaling station at Grytviken. He was soon followed by others and within a decade, six land stations and a floating factory at Godthul had been constructed by Norwegian and British whalers and the large-scale harvesting and processing of whales had begun. Humpbacks, fin whales, and the gigantic blue whales were the main species taken, together with smaller numbers of southern right whales, which were already rare. Between the opening of the first processing station in 1904 and the final closure of whaling here in 1965, 175,250 great whales had been rendered into whale oil and other products. The total harvest of marine mammals around South Georgia represents a slaughter which has never been matched elsewhere in the world. The whaling stations were substantial complexes with good accommodation, medical provision, and recreational facilities as well as the extensive whale rendering, storage, and transport buildings. The whale ships departed for the north with their cargos as autumn set in each year, but skeleton crews were left behind through the raw cold and dark of the winter months, to protect and maintain the base and its valuable equipment. These whalers were the first to give South Georgia a year-round human presence. The first residents were easily able to sustain themselves with plenty of fresh meat, not only with seal or whale meat and penguin eggs but also with introduced reindeer and numbers of free roaming pigs. Both species, but especially the reindeer, had a big impact on the vegetation of the island and thus also on the breeding birds. By the early years of the twentieth century man's heavy footprint on an erstwhile wildlife paradise, was firmly stamped.

Two other events merit passing mention in respect of the human history on the island. It was on the south-west side of the island in May 1916 that Sir Edward Shackleton landed with his five companions after a 16-day crossing of the Southern Ocean from Elephant Island. It was from here too that he, with Crean and Worsley, made

FIGURE 13 Grytviken.
© Palle Uhd Jepsen.

the epic crossing of the South Georgia mountains to reach safety at the whaling station at Stromness, before eventually planning the rescue of the remainder of his crew from Elephant Island. The other notable event was the short-lived, illegal occupation of the island by Argentinean forces in 1982 which was the precursor to the Falklands War.

The graphic beauty of South Georgia lies in the wild, snow-covered majesty of its mountain landscapes. The low grounds, jaw dropping as they may be for their prolific wildlife, do not merit the same order of beauty. However, as the snows of winter recede, these low coastal areas reveal a refreshing cloak of green. These are the robust plants of tussock grass. It is the dominant plant on the island, growing as much as 6 ft (2 m) high and half that in diameter and forms dense jungles where it has not been flattened by armies of seals. Amongst their roots, the tussock provides burrows for nesting petrels and shelter for breeding albatrosses, ducks, and pipits.

The wide, waving crowns of the tussock are favoured grazing for the reindeer and therein lies one of South Georgia's environmental catastrophes. Twenty-five reindeer were introduced by the Norwegian whalers in 1909 to provide fresh meat while later, up to the time of the Falklands War, their descendants were shot recreationally. However, inevitably, their numbers increased to reach in excess of 3,000 by the twenty-first century. They have had so serious an impact on the vegetation, especially the crucial tussock grass, that a decision has had to be taken to remove them completely from the island as part of a dynamic programme to restore South Georgia's natural environment.

On more open ground there are paler green lawns of Antarctic hairgrass with waving frothy heads. There are also unexpectedly colourful eye-catching carpets of russet: the stalked flowering heads of greater burnet. These patches of shorter vegetation are attractive feeding areas for the tough little South Georgia pipits, the only song bird on the island. But this is not a land of sunshine and blue skies. It lies in the track of the depressions constantly rolling through from the west and is characterized by weather that is predominantly wet, cold, and windy. A pall of cloud overhangs the island much of the time and, even in summer, occasional desiccating föhn winds, sweeping down from the mountains, produce brief but violent events. At other times bitter winds, short-lived but numbing, arise suddenly and accelerate over the mountains and down from the glaciers.

However, these are the conditions on South Georgia in which the multitude of birds and mammals thrive, fuelled by the great bounty of the sea. To visit one of the breeding beaches in the austral summer is to experience one of the greatest concentrations of wildlife on the planet. Wide, expansive beaches on the north side, such as those at Gold Harbour, Salisbury Plain, and St Andrews become a virtual carpet of penguins and seals. King penguin numbers have burgeoned in the last century and exponentially so in the past two

FIGURE 14 Elephant seal.

© Palle Uhd Jepsen.

decades. On St Andrews Bay the colony is now over 150,000 pairs and still increasing. In this bay alone there are also some 6,000 breeding elephant seals and a huge number of fur seals.

To have the privilege of stepping onto a beach such as this, in a moment of time, is an unforgettable experience. The elephant seals lie all over each other, huge wallowing mounds of blubber, stinking foully from both ends, for they break wind, belch, roar, and snort as they continuously try to rid themselves of armies of nasal mites and other parasites. The males are enormous, up to 4 tons in weight and as much as eight or ten times heavier than the females. On land they do little other than roll in their mud wallows, moult, give birth, and mate again. At sea they are transformed into masters of the ocean, spending almost all their time underwater, surfacing occasionally to breathe, feeding on a diet of squid and fish, and capable of diving to almost 3,000 ft (900 m) to find them. Fur seals are scattered more randomly. They lie out above the tideline where the resident bulls assemble their harems and where the females give birth shortly after landing and then mate again a week or so later. There is nothing handsome about the fur seals as they too lounge on the beaches; their open-mouth threats display an evil-looking pair of yellow incisors. On the same beaches the king penguins breed cheek by jowl with each other in dense throngs. There is an ever moving passage of penguins to and from the sea and the colony. The incoming birds surf in on the breakers, scrabble to stand upright, and then march inland with chests puffed out and heads held high with evidently regained dignity. Either individually, or in Indian file, they march purposefully, weaving through the fur seals when necessary and making their way back to their waiting young. In their first months the young are warmly clothed in thick down suits and as the season progresses, thousands of them huddle together in dense, warmth-hugging crèches. They are dark tan in colour, earning them the old sailors' name of 'oakum boys' as they reminded the sailors of the boys who became coated with the tarry oakum they hammered between ships' timbers to waterproof them. Other opportunists constantly patrol the beaches. The strange white-plumaged snowy sheathbills—odd, chicken-like birds—meander around the penguins and seals to profit from fur seal carrion and scavenge waste, however disgusting it may be. Brown skuas and grotesque southern and northern giant petrels prey on eggs and later snatch careless chicks, while kelp gulls forage the beaches and occasional leopard seals, straying from further south, police the surf for unwary penguins. All this effervescence of life, this explosion of nature's profligacy, defines the day shift. After dark it is supplemented by the dark shadows of innumerable seabirds—white-chinned petrels, diving petrels, tiny Wilson's storm petrels, and others in their millions, returning or departing from the nesting burrows in the tussock.

The rich waters around South Georgia are not exploited solely by the wildlife, for there is an active and strictly regulated fishery administered from King Edward Point at Grytviken. The shallow seas are protected by a 200-mile (320-km) zone around the island

and its satellites, the South Sandwich Islands and Shag Rocks. The fishery is extremely valuable, worth some £3 million per year in revenue from licences issued by the South Georgia Government, most of the revenue being ploughed back into fisheries protection and research. The research is centred in the laboratories at King Edward Point where a team of scientists and support staff work during the summer, reducing to a skeleton of four in winter, including a doctor. On Bird Island, off the north-west tip of South Georgia, there is also a small research station where work is concentrated on fur seals and seabirds, particularly albatrosses. Fifteen thousand pairs of four species of albatross breed on the island, although numbers are declining drastically, mainly through deaths far out at sea associated with long line fishing. Other species—penguins and burrowing petrels—also breed in prolific numbers and make this small island one of the most densely populated seabird sites in the world.

Bird Island is mercifully rat free but the situation on the main island is very different. Here, as on other oceanic islands around the world, rats have been present for the past 200 years. The arrival of man was accompanied by the inevitable landings of rats from infestations on his ships. The precise size of the rat population on South Georgia can never be calculated, but it certainly runs into millions. Over the years the rodents have made huge inroads into the populations of many of the seabirds, especially those smaller species nesting in burrows among the tussocks. Also at risk are two land species, the South Georgia pintail and the South Georgia pipit. Whaling stations were established in seven different locations which mean that there were seven original nuclei for rat colonization. However, the spread of infestation was even wider than this, as sealers landed on many sites and left the rodents behind them. In some cases such populations remain isolated because of tide-water glaciers separating a site from others on the same coast, but this was not always the case as, for example, in Stromness Bay where three whaling stations were mutually accessible.

In 2011 an eradication programme was initiated with the aim of the total extermination of rats from the island. Such work has been successfully undertaken on numerous islands in recent decades but South Georgia is a far more serious challenge, not simply on account of its size, daunting though this is. The island is some ten times larger than any other on which rodent eradication has been attempted and, to be successful, every last rat has to be killed; any individuals missed render such an operation a failure as the population can re-establish very quickly. Millions of pellets of brodifacoum are distributed by helicopter in affected areas. Urgency is added to the issue in the knowledge that as some of the glaciers retreat, the existing natural barriers between infected sites will disappear. In March 2011 initial control was undertaken and a successful first phase resulted in an extraordinary 29,650 acres (12,500 hectares) being declared rat-free. No government funding has supported the project and over £2 million was raised initially, towards the estimated £7.5 million needed to complete 198,000 acres (80,000 ha) by 2015.

So, man's relationship with the wildlife of South Georgia has gone full circle. It is an island that, although still supporting enormous populations of birds and sea mammals, has seen the devastation of some of its species on an unprecedented scale. After almost two centuries of wholesale exploitation since it was discovered by Captain Cook, the overriding emphasis now is on protection, conservation, and research. There is also limited access for tourists to experience the wonders of the island. Tourism thereby makes some contribution to the island's economy, although the amount raised is not large compared with the fishery income which accrues from one of the best regulated fisheries in the world.

A steady stream of cruise ships discharges passengers at Grytviken in the austral summer and many from the smaller ships make landings on specified beaches, under strict conditions and control. Landings at Grytviken give the opportunity to visit the post office and museum and most people make the short walk past elephant seal wallows, resting fur seals, South Georgia pintails, and groups of king penguins, to visit the little cemetery and pay homage at Shackleton's grave. Much of the extensive building that formed the whaling station at Grytviken has been demolished for safety (none of the other South Georgia whaling stations is open to visitors) but the whalers' Norwegian church is perfectly maintained and still operational. Man's impact on the island wildlife in the past has been severe but, in the twenty-first century, there is a real possibility of seeing serious restoration undertaken with the long-term prospect of a major recovery of its unique native communities of vegetation, birds, and mammals.

Halfmoon Island

Halfmoon is barren, remote, and has an almost tangible feeling of cold eeriness to it. It is an isolated, low-slung shoulder of rock at the south-east corner of the Svalbard archipelago. This is no Arctic oasis, unlike so many islands in the northern polar seas. It is the definition of desolation and soullessness. With virtually no soil, it supports only the most meagre vegetation and the sparsest range of birds and mammals. However, there is a strange magnetic quality about it and I love to walk the island in the brief, cheerless summers it experiences. On a still day, away from any companions, one can fantasize about being the only person on the island, and it is then that the silence is captivating. In our busy world, it is rare indeed to be free from any sounds, however distant. On Halfmoon, on such a day, stand away from the soft lapping of the ice-choked waters on the shoreline and listen to the silence. There is no sound anywhere yet you can 'hear' the silence in your ears. It is a large part of Halfmoon's alluring magic.

The Svalbard archipelago lies far out in the Barents Sea, 400 miles (640 km) beyond the North Cape of Norway, more than 600 miles (960 km) inside the Arctic Circle and deep within the latitudes of the High Arctic. Notwithstanding its geographical position—its northern headlands are a mere 500 miles (800 km) from the Pole—it is more accessible than any other Arctic lands at the same latitude. The simple reason for this is that it is blessed by the warm waters of the northern arm of the Gulf Stream which sweeps past the British Isles and the Faeroe Islands to bathe the western coast of Spitsbergen and keep that side of the archipelago virtually ice-free most of the year. In brochures and in casual conversation the archipelago is frequently referred to as 'Spitsbergen' instead of 'Svalbard', although the former is specifically the name of the western island of the group, the largest one and the only one with a permanent human population (excepting the meteorological station on the distant island of Hopen at the southern end of the archipelago). Because of its relative accessibility, Svalbard supports a reasonably thriving tourist industry, mainly in the form of cruise ships bringing passengers during the perpetual daylight of the summer months to experience this remote Arctic region, its majestic landscapes, and its glorious wealth of wildlife.

Contrary to the general belief that, whatever the season, these High Arctic regions suffer bone-chilling cold and insufferable weather and are unsuitable for human existence at any time of year, this is far from the truth. Needless to say, there are days in summer when katabatic winds slice off the glaciers and sleet showers can almost cut your face, but these are rare. I have been to the Arctic many times and to Svalbard in particular, and have enjoyed countless days of brilliant sunshine, calm seas, and pleasant warmth. In summer the sea cliffs, the inland lakes, and coastal pastures are home to countless thousands of breeding birds and concentrations of large mammals. Many areas are carpeted with an exotic, multi-coloured tapestry of wild flowers. For a short season colour is everywhere, in the brightness of flowers or the subtleness of pastel colours of mosses and lichens. Reindeer—the short-legged, sturdy Svalbard breed—graze many of the coastal flats, behind which are the jagged outlines of the mountains that originally gave Spitsbergen its name. Glaciers carve deep valleys at the head of each fjord as they slide off the mountains and break their backs with tumultuous sound and force when they reach the tideline.

But of course, this is the Arctic and conditions elsewhere are not like those enjoyed on the western side of Spitsbergen, which benefits so much from the gift of the Gulf Stream's current. Travel round the north of the island and penetrate the narrow Hindlopen Strait on the east side, with the huge ice cap of Nordaustlandet to port, and the world is a different place. No soft Gulf Stream here, for the shifting pack ice frequently log-jams the Strait for mile after mile. Only ships with ice-hardened hulls can penetrate and in some years even they cannot pass. Here is the land of the fragile-looking, but tough, ivory gull, one of the few Arctic birds that forgo the opportunity to abandon these Arctic wastes in winter. The shifting flows of ice in summer are home to walrus, polar bear, and its primary food the ringed seal. At the south end of the Strait is another ice-bound island, Edgeoya, with its satellite, Halfmoon Island, two miles offshore, locked together by landfast ice for most of the year.

Even in summer, when the winter snow has cleared the ground, it is difficult to think of more forlorn landscapes than those found in this area. The heaving sea is a heavy grey, the skies are leaden much of the time, and thick clouds hang low, often sinking even lower until fog obliterates everything beyond 20 yards (18 m). When there is visibility, the great icecap on Edgeoya dominates the skyline to the north where it merges invisibly into the pallor of the sky. The whole surface of Halfmoon speaks desolation.

Is it possible that such a place can ever claim the sort of raw, primal beauty to be found in many of the earth's wild places? Most of the island is bare rock, much of it covered in a skating rink of *umbilicaria* lichen, lethally ankle-wrenching and potential leg-breaking to walk over when it is wet, which is all the time in the summer melt. No garden of Arctic plants carpets the ground to bring lightness and colour. Only here and there are shallow, sheltered pockets of soil sprouting drifts of pale Svalbard poppies, their yellow parabolas turned towards the hopefulness of sun, to glean the extra few degrees of warmth. Here and there in the crevices between rocks there are tiny clumps of tufted saxifrage and a few bog saxifrages which grow in wet hollows. Offshore eider ducks rise and fall on the swell and Arctic skuas and long-tailed skuas drift over the island. On one of the dark, sterile pools a melancholy pair of red-throated divers is usually to be found. They nest among the pebbles and flotsam on the margin of the pool and fly the few yards to the sea to bring small fish for their two chicks, bobbing on the rippled surface. Hardy snow buntings, most northerly of all passerine birds, flit amongst the bare stones. The overwhelming impression the place transmits is of a soulless lunar wasteland, a place that God forgot. An enforced stay on this island would be to inhabit a Hell on Earth. It is a place at the end of the world that can surely have nothing whatsoever to offer man.

And yet, improbably, even in this frozen wasteland, man's footprint is everywhere and his inexorable quest for its wildlife has brought many Arctic hunters here over the centuries. The most cursory walk away from the shoreline reveals uneasy signs of his former visits. There are walrus in the area, plenty of Arctic foxes too and it is on one of the main routes for polar bears moving to and from denning areas on Hopen Island or King Karl's Land, across the sea to the south. The ivory of the walrus' tusks and the skins of fox and polar bear were very valuable commodities and man, the resolute hunter, could tolerate even the desolation of Halfmoon in return for the rich and profitable hunting it provided.

It is not known at what time in the distant past the first Siberian hunters visited Halfmoon, but it was probably far, far back in time. Bleached whale bones lie in parts of the island testifying to past activity and some of these, intriguingly, are not near the shore but in the centre of the island and are thought to be very ancient. It is likely that whalers discovered Halfmoon in the early 1600s as an island is shown south-east of Edgeoya on a Muscovy Company map of 1625. Russian hunters were certainly regular visitors from the late seventeenth century, followed by Norwegians a hundred years later, and Norwegians were the first to succeed in over-wintering in this desolate place in 1820. Hunting continued well into the twentieth century, until the area was declared a sanctuary in 1973. Wherever you walk on the island there is evidence of the trapping and killing that took place, mainly in the form of relict fox and bear traps. Heaven alone knows how many of these lethal devices there were at the height of hunting, certainly scores and many of the stony eminences are crowned with the visible remains of fox traps. The wooden framework of the trap was some 2 ft (60 cm) square, set on the ground with a heavy stone as a fitting 'lid', propped up and held there by a trigger stick to which the bait—often a Ptarmigan head— was attached. As the fox took the bait and displaced the trigger, the lid fell crushing the animal with the weight of the heavy stone but not damaging the valuable skin, as shooting would do.

In the latter phases of hunting the most popular method of taking the dangerous polar bears was also an ingeniously simple but lethally efficient trap. It comprised a wooden box, raised on legs so that it was at the comfortable height of a bear's head. One end of the box was open and the other closed except for an opening large enough to take the end of a shortened gun barrel. Bait was placed in the box and wired to the gun's trigger so that as the bait was seized the trigger was pulled and the bear took the full force of the charge face-on. Halfmoon is liberally scattered with the remains of these deadly devices, often accompanied by scattered skeletal evidence of the victims, for it was only the skin that was the valuable commodity. The Norwegian hunter Henry Rudi killed 117 bears himself in the winter of 1937–38. As late as 1970–71, Per Johnson over-wintered alone on Halfmoon and took a full permitted quota of twenty-five bears. (He also admitted to taking 140 nearby on Edgeoya in the previous winter when there was no quota.)

FIGURE 15 Polar bear trap.

© Susan Barr.

There are other, more pitiful signs of man's time on the island in the form of the graves of those for whom the challenge was too great and the conditions too severe. Apart from several single graves here and there, barely discernable among stone piles, there is one group covering the mortal remains of twelve hunters who failed to survive, out of a party of eighteen, from the last Russian over-wintering party in 1851–52. The low mounds of rocks covering the bodies lie close by the shoreline in the arc of the bay facing Edgeoya on the northern shore of Halfmoon, which gives the island its name. A more forbidding, forsaken, and desolate spot for a final resting place is hard to imagine.

Among all the human tragedy that was so clearly played out alongside the plundering of wildlife on Halfmoon, there remains one story of human survival on the island that, even after 200 years, still seems to defy belief. There are endless stories of human survival against overwhelming odds in polar regions, but none, I believe, that matches the remarkable, but little known, story of four Russians on Halfmoon Island in the eighteenth century.

In May 1743 a small wooden *kotch* sailed from the port of Mezen in northern Russia with fourteen Pomor walrus hunters on board, destined for a summer in the relatively safe waters of the west coast of Spitsbergen. After some two weeks of sailing westwards they hit bad weather in the Barents Sea and were pushed north by strong winds until they became inexorably trapped in moving pack ice off the southern coast of Edgeoya. Their fragile craft, fine on the open sea, was ill-matched for the conditions they now encountered and they were in serious danger of being crushed among the shifting, grinding ice flows. Drastic action was needed. Aleksei Inkov, the pilot, had not been to the area before but, through hearsay from other Mezen sailors, he believed he had some idea of where they were and that on the nearby island (it would undoubtedly have been Halfmoon Island) there was a hut that had been erected earlier by other Pomors. He volunteered to cross the moving ice to attempt to locate it as an emergency refuge for the crew, should the need arise and the ship founder. Three colleagues volunteered to go with him. The four set off, taking with them only token supplies for an overnight stay. These amounted to one musket with twelve balls and charges of powder, 20 lbs (900 gm) of flour, an axe, a tinder box, a kettle, and a knife. Each man had his wooden pipe and a pouch of tobacco. They made their way across 2 miles (3.2 km) of treacherous, shifting ice and, reaching the frozen land safely, remarkably found the hut, just as Inkov had guessed, on the far side of the island. In this forbidding and frozen place they rested overnight and in the morning retraced their steps over the rocks to the far shore. One can only imagine the feelings of hopeless desperation and shock when they reached the spot from where they should have seen the ship out across the ice, to find absolutely nothing and slowly realize that during the night it had been crushed in the ice and vanished from sight, together with their ten colleagues.

The four castaways were thus marooned on Halfmoon Island without resources and were to all intents condemned men. However, in the most astonishing feat of survival, with no food or tools, save the few pitiful items they had carried with them, they began what was to be over six years of isolation in one of the most challenging environments on earth.

They repaired the hut, wind-proofing it with moss where necessary. With their twelve musket balls they soon killed twelve reindeer which lasted them until the first autumn. Showing remarkable native ingenuity, they made spears and lances out of the abundant driftwood and the nails that some pieces contained. With these lances they fought by hand, and killed, a total of ten polar bears during their time there, without any of them suffering injuries. They made bows from the roots of driftwood larch trees and polar bear tendons and across the years killed some 250 reindeer which migrate across the ice from one island to another. Throughout the duration of their isolation there was no vegetation or other food and they ate nothing but polar bear, Arctic fox, and reindeer, successfully avoiding scurvy through their own belief in taking plenty of exercise and drinking fresh reindeer blood. With clay from a small hollow on the island they baked oil lamps, fuelled with reindeer fat, and with their tinder box and flints they made fire. Fire was so crucial to them for warmth and light in the total darkness of deep winter that they succeeded in keeping it alight (for they had no way of relighting it should they lose it) for the entire six years! Throughout the summer months every year they watched, forlornly, for any sign of a ship and the rest of the year they somehow endured the loneliness and the bitter depth of winter cold and darkness.

It was in August 1749 that they eventually saw a distant ship, lit fires, and waved reindeer-skin flags. Shortly before this Fedor Verigin, the oldest of the four, had passed away after months of illness and been buried outside the hut. The remaining three men were taken aboard and eventually made their way back to the port of Arkangel, almost literally having returned from the dead after six and a half years.

I have been to Halfmoon Island in summer several times and each time it makes the same impression on me. It is God-forsaken beyond words and produces a brief shudder, but at the same time it has a strange allure. The pale form of a glaucous gull, drifting, veiled through the mist over the barren waste, is a ghostly reminder of so much tragedy that has happened here—both human and wildlife. Whatever the feeling, I would return, again and again.

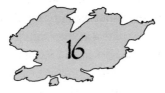

Great Skellig
(Skellig Michael)

30 May 1966: light southerly breeze, gentle swell and 3/10 cloud.

The Commissioners of Irish Lights' tender, *Ierne*, lay off Blind Man's Cove on the east side of Skellig Michael. It was as calm a day as one could hope for on the open ocean off the south-west corner of Ireland—as indeed it needed to be to effect a landing on this daunting rock. A cutter was lowered and navigated carefully alongside the beetling cliffs where it rose and fell on the gentle swell, its position skilfully held by the strong arms of the four oarsmen. The arm of a derrick swung out from 100 ft (30 m) or more above the sea and lowered down an elemental Bo'sun's chair—a stout cross piece of 'broomstick' at the end of the rope. If I had a moment's hesitation about this way of landing on the rock for a week's stay, it was simply that I was joined on the chair for the vertical journey by a hefty lighthouse keeper and his gear. On a difficult operation therefore, while good conditions and a sturdy rope prevail, they took two of us up for the price of one.

This was many years ago and was my first of several subsequent visits over the years to what is unquestionably the most extraordinary island on which I have ever had the good fortune to spend time. Skellig Michael is, of course, still as gloriously wild, precipitous, and awe inspiring as it was all those years ago, but in those days there were only exceptional visitors on a day trip from Valencia Island or Cahirciveen. Now the lighthouse has long since been automatic (1980s), and servicing the light is done by helicopter and the increasing demand for visiting this remarkable UNESCO World Heritage Site has resulted in access being strictly limited. Modern progress may have laid its finger on this western outpost of Europe, but its ineffable mystique remains ineradicable.

Skellig Michael (Great Skellig) is a truly unique and quite remarkable site. It boasts a spellbinding monument to human endeavour and determination while simultaneously supporting a great multitude of avian residents. Like its smaller sister, Little Skellig, a mile away, it rises absolutely sheer out of the ocean as an unscalable cliff-girt pyramid of rock, to a height of 714 ft (217 m). The wide Atlantic lashes furiously at its feet most days of the year. On one winter occasion waves struck with such force that they broke the thick glass of the lighthouse lantern, 175 ft (53 m) above the sea. Sea conditions, certainly between November and April, frequently render it wholly unapproachable by sea, and yet this forbidding rock was the home to Celtic ascetics as long ago as the sixth century AD. The UNESCO citation described it in 1996—perfectly accurately but somewhat prosaically—as 'a unique religious settlement on a pyramidal rock in mid ocean which illustrates the extremes of Christian monasticism'. Extreme it is in every respect. There are twin peaks. The highest is a rock climbing challenge and was a principal penitential station wherein the penitents scaled the nerve-racking rocks, past the Station of the Cross, squeezed through the 'Eye of the Needle' just below the summit before edging along an overhanging rock with a dizzying drop 700 ft (210 m) below, to kiss the stone carving at the tip. It is a fact that the retreating descent is equally as daunting as the ascent.

The northern summit is 100 ft (30 m) lower and considerably broader although still edged with stupendous cliffs lined with breeding fulmars. It is the pulse of the historical and archaeological heartbeat of the island. Here, clustered together within a dry-stone walled compound, are the six corbelled bee-hive cells of the monastery, still in almost perfect condition after 1,400 years. The shells of two small oratories are close by and there are also the remains of a twelfth-century, mediaeval chapel within the compound. The vacant east window of the chapel perfectly frames the sister island of Little Skellig with its immense gannetry. The whole surface of that jagged rock is whitewashed with the mass of nesting birds and

FIGURE 16 Gannets on Little Skellig.

© The author.

their droppings, while the air above and around is a blizzard of snow-white, black-tipped gannets. In the evenings, long Indian files of birds, returning from fishing excursions, trace their lines across the sky.

At the site of the seventh-century monastery one can only stand in awe—it *is* one of the most remarkable of Christian sites anywhere—not simply marvelling at its almost celestial location and panoramic outlook, but recoiling at the scale of building involved in the vertiginous retaining walls and wondering at the problems of subsistence that challenged even the frugal requirements of the ascetics. This was life at the very edge. It is believed that the eremitic monastic community comprised probably no more than a dozen monks together with their abbot. How then did they subsist in this remote and isolated sanctuary? They certainly had a small terraced garden or gardens, supported by enormous retaining walls on the sheer cliff slopes, although these small areas could never supply even the frugal needs of the monks throughout the year. It is reasonable to assume that the ready supply of meat and (seasonally) eggs provided by the huge populations of seabirds on the island, were harvested. We can presume too, that they had skin-covered boats, similar to the currachs on the western mainland and the Blasket Islands 20 miles (32 km) to the north, which allowed them to fish when the weather was clement. Furthermore, a mile or so across the sea, the Little Skellig, precipitous though it also is, was within easy reach. It is one of the world's great gannetries and it is known for certain that in recent centuries birds were taken there, often in great numbers. Whether the gannets bred there and, if so, in what numbers 1,300 or 1,400 years ago is not known, but with or without gannets the island would certainly have yielded them fresh seabird meat in one form or another. Unsurprisingly there is a history of harvesting seabirds in the area, although how far back it stretches we can only guess. Up to the late part of the nineteenth century, birds were collected by mainland parties, both for the value of their feathers and for the meat of young gannets. Earlier in the same century, the owners of Little Skellig went so far as to station guards in a boat offshore during the appropriate season to prevent unauthorized plunder, but even this did not deter some, such was the premium put on the birds. Lavelle (1976) relates an incident when men from Dunquin plundered the birds but were intercepted by the guards, resulting in a ferocious confrontation and two fatalities among the guards before the Dunquin men successfully fled with their booty. Puffins were taken in large numbers on Great Skellig up to the early years of the twentieth century, both for their feathers and for salting down for the winter. Rabbits were introduced to Great Skellig in recent centuries and are now numerous, although they were of course totally unknown in Ireland long after the monks had left. Fulmars, important food items on places like St Kilda, did not arrive here until 1913.

The ancient routes up the rock from sea level were challenges in themselves. There were three paths, with stairways hewn into the solid cliffs. One (the lower parts of which were blown away 150 years ago during the construction of a lighthouse jetty) ascends from Blind Man's Cove on the east side, another from Blue Cove on the north side of the island and a third from the south side, near Cross Cove.

The lighthouse engineers built the tiny jetty in Blind Man's Cove on the north-east corner of the rock (useable for any purpose only in the most favourable of conditions) in the 1820s and from it, constructed a seemingly impossible walled road, literally on the cliff edge from there to the lighthouse at the south end. Half way along, and thus well away from the lighthouse, they built a stone shed for storing the explosive fog signals. In 1936 an electrical fault accidentally detonated the entire stock of 300 signals with an explosion 'that nearly lifted The Skellig out of the sea' (Lavelle 1976: 61). Close to this same spot, the lighthouse road cuts across the original monk's climb from the south landing, where there is now a colony of kittiwakes. It is from here that the modern access to the monastery is gained. A steep flight of some 640 steps climbs up through one of the few vegetated slopes on the island and leads up to a col between the two peaks, aptly called 'Christ's Saddle'. From here it continues up a narrow, steepling staircase of stone-built steps, with a fearsome drop on the left, leading to the short tunnel gateway through the walls of the compound and into the monastery.

Great Skellig certainly retains a powerful atmosphere. It pervades an ambience that I have never experienced as strongly on any other island. Predictably, this consciousness is strongest at the monastery but it prevails throughout the island, for the ascetic's footprint is everywhere, in improbable staircases, penitents' routes, rock markings, the remains of a cell clinging to an impossible ledge. Added to these facts the place is suffused with legends and myths. It was here that the Archangel Michael and the Heavenly Host visited the wild, sea-lashed rock with St Patrick to rid Ireland of snakes and other undesirable creatures. In even earlier centuries the legends provide other stories of natural and supernatural events related to this far-flung place. Here magic forces caused the wrecking of Melisius' ship and the loss of his two sons around 1400 BC, while Duagh, King of the World, rested here in 200 AD before his battle of a year and a day against Finn MacCool at Ventry. Whatever our acceptance of the richness of the unwritten chronicle of the past, there is no doubting the equally compelling factual history of man's association with the rock, and the recognizable aura of spiritualism that permeates Skellig Michael to this day. 'Reverential' is perhaps the word that best fits my feelings when on the rock. There is a sense of sanctity that always surrounds the place, despite the journeys through time and the less sensitive activities of nineteenth- and twentieth-century man. Skellig is special and doesn't let you forget it.

We do not know for certain when the monks finally abandoned the island, although it was probably in the twelfth century when they moved to Ballinskelligs on the mainland. However, in between times they suffered raids by the Vikings in the ninth century. They suffered plunder and murder on several occasions, first in 812, then eleven years later, in 823, and again subsequently. Nonetheless the community persisted and regrouped each time until the final withdrawal. The ancient monastery later supported medieval religious activity from the twelfth century, as witnessed by the chapel remains. 'Modern' man inhabited Skellig through generations of traditional lighthouse families from the 1820s to 1980s, but now they have gone and it is deserted again, apart from the intermittent trickle of day visitors on the limited calm days of summer.

Although there may be no permanent human dwellers now, Skellig Michael is far from being devoid of life. Apart from the 27,000 pairs of gannets on nearby Little Skellig, the Great Skellig is one of the most important seabird fortresses off the western coasts of Britain. During daytime in the summer months, the vertiginous slopes alongside the steep path to Christ's Saddle are populated by throngs of puffins and cliff ledges above the sea, all round the island, support tenements of fulmars, guillemots, and kittiwakes, with pairs of razorbills among rock crevices and hollows in the cliffs. A solitary pair of scimitar-billed and crimson-legged choughs, aerial masters of the eddying winds and updraughts, lives on the wild slopes.

However it is during the short nights of summer that the island truly comes to life. Dark moonless hours spent at the monastery— and I have been privileged to enjoy many of them—require a strong nerve, at least initially, and an appreciation and understanding of the night-time inhabitants. A climb by torchlight up the staircase from Christ's Saddle in the fading dimness of dusk gives the first clues, as a soft purring comes from under almost every step. Deep within, a storm petrel, smallest of all seabirds, home from the sea, sits on a single white egg awaiting the return of its mate. Bend down close to the step and you may detect a whiff of the characteristic musty odour of the little birds themselves. These incidental encounters are only the beginning. As darkness intensifies on a moonless night, the air is suddenly rent by a spine-chilling, demoniac caterwauling. In no time there is another and another and the night-time chorus of returning Manx shearwaters has begun. The darkened sky is full of noise as the birds call to contact their mates in underground burrows in the daytime puffin slopes and elsewhere. Amongst the confused chorus of identical raucous calling, each bird in its subterranean nest recognizes the individual call of its mate. There are several thousand pairs that breed here; at one time a pair laid their egg and reared their single chick each year on the soft earth in Cimmerian blackness in the centre of one of the bee-hive cells. Despite the trickle of inquisitive visitors, one resolute pair still tenants the inside of one of the cells. Through the hours of darkness and until the first glimmer of light in the east, this wild witches' laughter continues and occasionally the dark shadow of a bird flashes close by.

Around the monastery, through the night, it is not only the shearwaters that populate the air, for there is endless traffic as thousands of the diminutive storm petrels come and go to and from their nests. These petrels nest sometimes in earthy burrows, but here on Skellig Michael they have little need of the soft earth because the vast expanse of dry stone walling offers an endless choice of nesting crevices. In this respect the monks have been responsible for helping to increase the population of these birds on the island. The air is alive with their trembling wings as they hover and stall in front of the walls in the torchlight beam. The colony here is of many thousands of birds and, as with the Manx shearwaters, they invisibly add a telling night-time element to the magic of this unique citadel.

How far west and out to sea are the Skellig Rocks? St Kilda, 40 miles (64 km) beyond the Outer Hebrides, usually lays claim to being the most westerly point of land in the British Isles but it is in fact a poor candidate for the claim, as the whole of western Ireland is farther west and the Skelligs, 9 miles (14 km) off the west coast of County Kerry, are a good 70 miles (112 km) further west than St Kilda.

However, even the Skelligs are not the remotest point, for that accolade belongs to Inishtearaght, 15 miles (24 km) to the north, the most westerly of the Blasket Islands and, at longitude 10° 40´, indisputably both the British Isles'—and Europe's—farthest outpost (save the wave-lashed stump of Rockall far out in the Atlantic). Like the Skelligs, Inishtearaght rises sheer from a tumult of sea but was never inhabited—nor could it have been—until the building of the lighthouse in 1870. Again it is a seabird metropolis with important colonies of fulmars, guillemots, puffins, and razorbills but it is most notable for its population of Manx shearwaters and a huge colony of storm petrels, possibly the largest in the world.

Ile aux Aigrettes

The Ile aux Aigrettes, lying in the vastness of the Indian Ocean, is no more than a tiny offshore satellite to the large island of Mauritius. It may be tiny, but its importance in the modern story of man and his relationship with wildlife in the Mascerene Islands is far out of proportion to its size. It is the slender thread that protects the remnants of a tropical wildlife paradise that has been wholly lost elsewhere. At no great risk of exaggeration, it can boast all the elements of a modern Noah's Ark and has set an international standard in the conservation of island wildlife where it has previously been ravaged by the hand of man.

Ile aux Aigrettes lies half a mile (800 m) off the south-east coast of Mauritius. When one is aware of the history of the island over the past couple of decades, it is easier to understand the tangible air of mystery and expectation among visitors on the short boat crossing. Unlike the jagged background of the main island which is volcanic in origin, Ile aux Aigrettes is composed of coralline limestone, overlaid on much of its surface by sand, blown soil, and the natural compost of ages: the decayed vegetation of millennia. The islet is indeed small, a mere 64 acres (26 hectares) in extent, roughly circular in shape and, with its coral origins, predictably low-lying. It is clothed in green forest right down to the water's edge, a jewelled emerald sitting in the dazzling turquoise waters of the tropical ocean. To appreciate the critical importance of this one isolated coral islet, it is necessary to understand the enormous damage that has been inflicted on the environment of Mauritius itself over the past 400 years and the catastrophic loss that much of its unique wildlife has suffered since its discovery by European sailors in the early sixteenth century.

Mauritius developed its own flora and fauna throughout the timeless ages of its isolation from other landmasses and, before Man's intervention, it supported a truly extraordinary range of species found nowhere else in the world. Once passing ships started to call here for revictualling or repair, immediate exploitation of the islands resources began. Forest was felled and some of the confiding birds and the large tortoises were taken for food, a practice that continued across succeeding eras. The island was first settled by the Dutch in the early seventeenth century and they began the process of forest clearance, but abandoned the island in 1710 having already overexploited the wildlife and left much of the surface deforested. Five years later the French claimed the island and by 1721 they had settled on it and continued the ecological devastation. Once the island was settled by Europeans, not only was the native vegetation and wildlife destroyed, but a veritable zoological roll call of alien animals was introduced—cats, rats, mongooses, crab-eating macaques, pigs, goats, Java deer, Indian wolf-snakes, Indian musk shrews, and mice. Several of these animals readily preyed on the island wildlife, which up to that time had known no land predators. To confuse the picture further, several non-native birds have also been introduced over the years, including Indian mynah, Madagascar fody, red-whiskered bulbul, zebra dove, ring-necked parakeet, and house crows. Mauritius became a biological chaos; many species sank to extinction and the island was flooded with exotic aliens. The pristine environment originally supported a rich treasury of endemic species of plants (at least 311 flowering ones strictly exclusive to Mauritius), reptiles, and birds (at least 22 species, of which 10 are now extinct[1]). Inexorably the forest cover on Mauritius was felled and, from the middle of the eighteenth century, great swathes of remaining parts of the island were cleared wholesale and planted with sugar cane. Only in the more mountainous south of the island is there still an isolated tract of original forest, including 25 square miles (65 km[2]) in the important Black River Gorges National Park. The combination of the comprehensive removal of native forest habitats over the majority of the island and the depredations of introduced mammals determined the inescapable demise of so many of the unique bird species and the complete loss of the former coastal forests.

The most famous and charismatic species to have been lost, as everyone is aware, is the dodo, a word almost synonymous with both Mauritius and the word 'extinction' itself. Although it remains the most famous extinct creature in the world, there was nothing very beautiful about the bird. It was a giant, corpulent, and rather grotesque relative of pigeons, standing almost 3ft (1 m) tall with adults weighing around 30 lbs (15 kg). Although it was completely flightless, it lived in an environment where no predators existed and was therefore totally fearless of man once he arrived. It was very numerous and the limitless source of fresh meat that it represented to passing mariners was too good to ignore. However, the belief that the harvesting of dodos by seamen was directly responsible for its rapid extermination is not, in fact, very close to the truth. It was certainly killed for the pot, but the meat was apparently tough and unsavoury and therefore universally fairly unpopular, especially at certain times of year depending on the food it had been eating. Other species on the island such as the flightless red rail and two giant tortoises were more favoured (both tortoises were exterminated, in common with three similar species on other Indian Ocean islands). The flightless red rail was a large bird, at least chicken-sized, and was easily caught. It was regarded as excellent eating and this certainly hastened its extinction within 100 years of its discovery.

The more critical factors in the dodo's demise were undoubtedly the twin effects of the wholesale destruction of the natural forest habitat in which it lived and upon which it depended for food, and the introduction of animals that preyed directly on the bird, all of which, particularly the pigs and macaques, preyed on the dodo and its eggs. Only 80 years after it was first seen, the dodo was extinct and the red rail followed a decade or two later.

Far removed as it is from other lands, Mauritius was a classic example of a rich and varied island biodiversity which had developed its unique communities over millennia. In these respects its history since the arrival of man is a tragic example of long-term ecological desecration. However, since the latter half of the twentieth century, huge efforts have been made to save and restore the populations of the remaining species that have teetered on the brink of disappearance and some of the results have been spectacularly successful. This is the point in time at which Ile aux Aigrettes assumes its importance.

The island has been put at the heart of efforts to restore natural habitats and re-establish lost or vulnerable species. Mauritius itself is still awash with introduced species and its size renders it almost impossible for eradication programmes to be undertaken in the foreseeable future. Ile aux Aigrettes is small, however, and although it too suffered serious damage and has been infested with many of the same introduced alien plants and mammals as the mainland, its size offers realistic possibilities for a gradual return of the ecosystem to its former state, whilst simultaneously offering a sanctuary for reintroduced plants and animals. The island was declared a nature reserve in 1965 and leased to the Mauritian Wildlife Foundation by the government in 1986.

The island certainly suffered great losses to its wldlife. Even as long ago as the eighteenth century, the large colony of dimorphic egrets (aigrettes) from which the island gained its enduring name, was eliminated. The island's forest cover was seriously degraded and rats, goats, and cats were released. Other species such as Indian musk shrew (or Asian house shrew), giant African landsnails, and Indian wolfsnake also found their way there and remain to this day. The Second World War saw the place taken over as a British military battery, peppered with gun emplacements and some of the remaining forest being cleared. At some stage a South American shrub, false acacia, was introduced to the island. It is a highly invasive plant, eliminating competitors and soon it covered most of the island; it is useful for goat grazing because it is so prolific and persistent but, as has been found in other parts of the world, disastrous in every other respect. The island was privately rented for a time and used for goat grazing in an attempt to control the plant but with only limited success. In 1985 an ambitious programme of environmental rehabilitation was launched through the Mauritian Wildlife Appeal Fund (later the Mauritian Wildlife Foundation). Its daunting aim is to restore ecosystems to their original state on Ile aux Aigrettes, eliminating all introduced flora and fauna and establishing a project for cultivating and replanting native trees and other plants, as well as serving as a refugium for the restoration of rare native birds and reptiles. The island contains many Mauritian endemic plants classified as Critically Endangered, for example bois de fer, bois de boeuf, bois de ronde, and the lowland coastal ebony. The best fragments of the original unique dry coastal forest are found on Ile aux Aigrettes, having been eliminated elsewhere round the coasts of Mauritius. Also significant is the exceptionally rare and very beautiful epiphytic orchid, *Oeniella polystachys*, the largest population in the world being found on this one tiny scrap of land. It has luckily survived the ravages of the past and is an exquisite forest plant, epiphytic on small trees in hot, dry locations, and displaying a long inflorescence of delicate white flowers.

Through hugely impressive labour, the majority of the entire island has been successfully cleared of the non-native vegetation and either replanted with native trees or allowed to self-regenerate, assisted by the fact that rats were removed from the island by 1991. The rats were successfully eradicated through trapping and the use of selective poisons, while other mammals were trapped and removed at the same time. The rats had been responsible for feeding heavily on the fruits and seeds of several of the endemic trees, thereby preventing natural regeneration. Since the eradication of rats, native tree seedlings have started to appear including a large number of the rare ebony seedlings. A wide range of endemic plants has been reintroduced to the island and a plant nursery established which initiated propagation and initially produced some 40,000 plants each year for replanting. Now that the re-planting rate is reduced, the annual output is still around 4–5,000 seedlings. The re-establishment of native vegetation and the control of alien animals is the first step in restoring the wider ecosystem and, although there are no native mammals other than bats on Mauritius, there are various important reptiles and some extremely rare emblematic birds to consider. One of the most striking reptiles is the ornate day gecko. It is a small creature, no more than 5 inches (12 cm) in length but startling in appearance. It comes in a dazzling variety of electric blues and greens, heavily patterned with red spots on its back and a brilliant turquoise tail with red bars. As its name suggests, it is a daytime animal which feeds on insects and small invertebrates but also enjoys sweet nectar and pollen. In 2006 the first translocation of Telfair's skink were carried out. The animals were restricted to Round Island, another small nature reserve 16 miles (25 km) north of Mauritius, where they are still extremely numerous, although they have been lost elsewhere on the main island and other offshore islets in the past. This first attempt to establish a viable population on Ile aux Aigrettes was hampered by egg predation by the musk shrew ('le rat musquet'), although the shrews themselves (together with agamid lizards, land snails, and wolf-snakes) were depleted by the skinks. In 2010 a further 500 skinks, together with fifty Guenther's geckos were taken from Round Island and it is hoped that the increased population of skinks will further reduce and possibly eliminate, the problematic shrew. The shrew arrived in Mauritius less than 200 years ago and has proved a serious problem on Ile aux Aigrettes since the 1940s where, within its omnivorous diet, it will feed on the eggs and young of native lizards. At the other end of the size scale the most impressive reptile to be brought to the island is the giant tortoise. In 2000 twenty-two animals of the closely related species from Aldabra were introduced to replace those exterminated in the past. They are helping to control introduced grasses and other alien plants and at the same time dispersing seeds of native trees such as the ebonies. After 200 years giant tortoises are again living and breeding successfully on Ile aux Aigrettes.

Despite the outstanding success achieved in the elimination of alien species, the restoration of native forest and the recolonization of reptiles on Ile aux Aigrettes, it is the story of three bird species which has made the programme of recovery world famous.

The Mauritius kestrel probably evolved as long ago as the Pleistocene era, inhabiting the forests and preying on a variety of large

FIGURE 17 Tortoise.

© Vikash Tataya.

insects, small birds, and lizards. It is believed that it was originally scattered throughout the main island and was fairly common until the 1940s. It suffered initially through the removal of its forest habitat and predation by introduced mammals. However the *coup de grace* was the widespread use of DDT from the mid 1940s to the mid 1970s. By 1974 only four individuals were left, including a single breeding female, making it almost certainly the rarest bird in the world. The story of its survival and eventual increase is one of the most remarkable conservation successes anywhere. After initial attempts at captive breeding were unsuccessful, a new attempt was made in 1979 on Ile aux Aigrettes, led by an inspirational Welsh biologist, Carl Jones, and supported by the Durrell Wildlife Conservation Trust and also by Birdlife International. Freshly laid eggs were taken from the wild and breeding pairs encouraged to re-lay through the provision of supplementary food. Fertile eggs were hatched in incubators and resulting young successfully hacked back to the wild. Within five years or so the wild population was estimated to hold as many as eighty birds and is now put at a stable population of around 600 individuals. The majority of birds that had been released on Ile aux Aigrettes have subsequently left the islet, moving across to the main island where they are now found in the Bambous Mountain Range on the east coast, with sub-populations in the upland forests of the Black River Gorges National Park on the west coast. It is a story of the spectacular success of one man's efforts to pull a species back from the very brink of extinction.

This story of conservation success is paralleled by two further examples: the pink pigeon and echo parakeet. Both are species whose populations were at the very limits of survival. The pink pigeon is a plump bird, related to the extinct dodo, and an improbably bright pink. Although its head is paler than the rest of its body, it has bright pink feet, a pink bill, and a coral red ring round the eye. Formerly believed to have been common throughout Mauritius, its numbers fell to only 12–20 individuals in the 1970s with the destruction of forest habitat and predation, at which point it was the rarest pigeon in the world. Captive breeding began in the mid 1970s at Black River village and the first birds were released to the wild in 1988. The pigeons feed from trees—they sometimes even hang at the end of flexible branches to reach the fruits—as well as on the ground, searching for small fruits and seeds. A target of 500 birds in the wild is within reach, helped by the establishment of several sub-populations, including that on Ile aux Aigrettes. Many of the free-flying birds on the main island show a preference for nesting in introduced trees as these are often the majority available and also offer best protection against rats and monkeys.

The history of the echo parakeet is one of similar success. It is a medium-sized green parakeet, very similar to the closely related Indian ring-necked parakeet. From a population of no more than 20–25 birds, Carl Jones and his team have successfully increased the numbers of the Mauritian bird to an astonishing 500 or more. The story of the rescue of these three species is an outstanding example of what can be achieved by concerted conservation efforts, technical skill, and inspirational leadership.

On Ile aux Aigrettes the resonance of the work that has been achieved is everywhere to be seen. New forest growth with native trees and ground plants abounds, giant tortoises are readily encountered along the paths, Mauritian fodies and Mauritius olive white-eyes have been reintroduced and are thriving, and the improbable pink pigeons are easily observed on the island. The forest stirs to the soft sound of cooing pigeons and the reintroduced Telfair's skinks and Guenther's geckos can be seen on most visits. The island suffuses a vibrancy and excitement that goes with the evident success of a truly challenging programme of restoration and renewal. Ile aux Aigrettes may be tiny and remote, but its reputation is of world status and it is a truly exciting place where immense credit is due to the staff there. On no other island in the world has such a diverse programme of biological recovery been achieved with such success.

Solovetski Islands

It is a fairly safe bet that most of the people reading this book will not have heard of the Solovetski Islands previously. They are indeed little known and little visited, hidden in a distant corner of northern Europe, far off any normal mainline routes. However, the islands have cultural and architectural history of enormous significance. Moreover the environment there supports an enviably rich diversity of wildlife both on land and in the surrounding waters. This is the land of twenty-four hour sun in the summer and the waving multi-coloured curtains of the Aurora Borealis in the winter; a land which, in its season, is abuzz with countless birds and insects and bright with rocky meadows carpeted with wildflowers.

This remote archipelago, comprising six larger islands and dozens of smaller ones in the Russian north, lies deep in the icy waters of the White Sea, a bare 100 miles (160 km) south of the Arctic Circle and on its southern side guards the entrance to the wide expanses of Onega Bay. The land area of the islands extends to no more than 135 square miles (350km²), and they are generally low-lying, only reaching a modest height of 280 ft (86m) at the summit of the highest hill on Anzer Island. Access by air from Arkhangelsk is possible but the real way to approach Solovetski is across the wild sea routes that have been followed for thousands of years by early man, medieval traders, invaders, pilgrims and, more recently, by the agents and doomed victims of Stalin's purges. Across the cold grey waters, well before you can make out the crouching contour of the islands themselves, the magnificent outline of the medieval monastery soars above the far horizon. The dazzling skyline of spires, onion domes, cupolas, and pink granite walls crown one of Russia's greatest architectural treasures. In the early fifteenth century the first pious ascetics espoused this lonely and isolated place and founded their spartan cells here, on remote Solovetski. Over successive centuries the sanctuary slowly developed into one of the most important and famous religious centres in Russia. It was built with its massive walls close by the water's edge, so that on a calm day its breathtaking facades are mirrored to perfection in the quiet shallows of the harbour bay.

Nonetheless, this is a fairly unforgiving environment where the days are dark for half the year and much of the surrounding sea is frozen for just as long. The glaciers of distant ice ages have etched their indelible mark on the scoured rocks and have discarded innumerable boulders everywhere across the surface of the islands so that, in places, the native trees and plants struggle to gain tenuous footholds where they can. Stunted shore-line birches are bent in obeisance to the winds that frequently sweep the islands, while elsewhere, ash trees and spruces, insinuating their roots amongst the lichen-felted boulders, provide dark canopies above. The environment may appear unforgiving, but that does not disguise the fact that in the months of spring and early summer the woodlands ring with the songs of birds returning to breed in the bounty of these sub-Arctic lands. The boulder-strewn open areas are studded with flowering plants summer-long—fireweed, angelica, meadowsweet, and wood cranesbill.

Innumerable lakes, fish-rich and crystal-watered, surrounded by wooded tundra, are scattered across the islands. On Bolshoi Solovetsky, at nearly 10 square miles (25km^2) the largest of the islands, some of the lakes have been conjoined by short channels and canals, laboriously hand-dug by forgotten generations of monks and their workers, to form a lacework of interconnecting waterways. Boats to explore the waterways are now provided for the scatter of Russian and foreign tourists who make the long journey to the islands. From one of the lake-bank landings a winding path leads soundlessly through dark, silent woodlands and along a tall avenue of Norway spruce before emerging into the bright sunlight of an open clearing, backed by a faded nineteenth-century lodge, an ice house, and a chapel. This is the site of the monastery's botanical garden, nestling within the welcome warmth of a south-facing arc of low hills. In the brightness and quiet of the garden a willow warbler sings from the fringing trees and a spotted flycatcher makes short excursions from an apple tree close to the house; both these tiny birds are travellers from distant Africa, at the end of their annual migration to breed this far north. The botanical garden was created after the building of a hermitage here in 1822, developed by the monks, and used for the cultivation of a wide range of fruit trees and introduced flowering plants. It is a quiet and peaceful place, the antithesis of the forbidding ghostly shadows that haunt many of the other parts of Solovetski. The neglected garden is in the slow process of careful restoration after decades of twentieth-century neglect and decline.

Solovetski's fame reaches far back, principally to the foundations of its religious origins. The gigantic Kremlin with its magnificent cathedral, Peter the Great's church on Zayatzky Island, and a host of other ancient chapels, sketes, and hermitages are the indelible footprint of 'modern' man. However, these remote islands have been a sacred place, isolated and magical, for at least 7,000 years. Numerous Neolithic stone labyrinths and ancient burial mounds betray the presence of early man and are one of the defining characteristics of the islands. The circular labyrinths, intricate stone structures, are similar to many others round the coasts of northern Europe, but the

FIGURE 18 Labyrinth.

© The author.

thirty-five on Solovetski represent some of the best preserved.
On Bolshoi Zayatsky Island alone, there is one group of fourteen
labyrinths in an area less than an acre (0.4km²) in extent. They comprise spiral rows of boulders, one row within another, in the form of
a maze with a single opening. The purpose of these ancient structures is unclear, shrouded in mystery and, tantalizingly, lost in time.
It is suggested that they may have represented the borders of the
underworld or that they were used for rituals in helping souls journey from this world to the next. It is certainly likely that some had a
more mundane purpose, as fish traps. Started above the low water
mark, the 'mouth' of the labyrinth faced the sea to allow ingress of
the fish on the rising tide and enabled the fishermen to walk the captive catch gradually into the inner recesses of the complexes as the
tide ebbed and left them stranded. Such labyrinths, whatever their
function, are now high and dry above the shoreline since the land,
freed from the weight of hundreds of metres of ice, has 'rebounded'

and risen over the millennia since the last Ice Age. The mysterious labyrinths are now mostly clothed in mosses and lichens, crowberry, bilberry, and other low-growing plants, but their forms still remain as clear as they were several thousand years ago.

The great monastery was founded in 1429, standing on the narrow spit between the sea front to the west and the Holy Lake on the east side, which is the outflow of waters from the network of lakes and canals. The monastery rapidly became a magnet for pilgrims throughout Russia as its fame spread and its buildings expanded to accommodate both pilgrims and legions of volunteer labourers. Sustainable agriculture and regular fishing from the fertile waters in and around the islands, together with the harvest of birds and eggs, supported the community and fish also provided an important item of trade. But, the catalyst of Solovetski's enormous wealth was salt. High salt-content brine occurred nearby on the monastery's extensive mainland possessions and, with sufficient manpower, could be extracted and processed to provide the most valuable trading commodity of all. By the seventeenth century there were some fifty salt works in the region and the enormous profits which they generated enabled the development of the magnificent monastic buildings and formidable bastions that we see today. The skyline is an awesome spectacle of tall domes, minarets, bell towers, and cupolas within the massive pentagonal stone ramparts with their five huge corner towers.

Such defences were necessary because the monastery suffered numerous offensives across time; the protracted Swedish–Russian wars late in the seventeenth century impacted heavily on Solovetski and its mainland possessions and the English navy penetrated the White Sea and attacked the bastion unsuccessfully during the time of the Crimean War. The fortress even had to endure siege by the Russians themselves, during a rebellious dispute in the second half of the seventeenth century when the community was admonished by the Tsar over its refusal to accept liturgical and ritual reforms.

For all its pious orthodoxy and sanctity, the monastic settlement has suffered more than its fair share of conflict and trouble over the centuries.

There is a raw, sub-Arctic beauty about much of the landscape on the islands, which belies the lushness of some of the vegetation and a genuine wealth of wildlife. The archipelago benefits from a rare island microclimate which spares the islands from the worst of the harsh winters. As a result, they are notable for the rich flora that they support and the range of unexpected plants found in the botanical garden. The great variety of birdlife to be found here—some 190 species have been recorded—has resulted in international designations for their protection. In summertime seabirds abound and there are large colonies of Arctic terns, lesser black-backed and common gulls, and black guillemots. Common eiders, with their broods, enliven the shoreline here and there; in earlier centuries the White Sea area was a prime producer of eider down but its importance diminished and then evaporated in the twentieth century, particularly after the 1917 Revolution. Both black-throated and red-throated divers occur on the multitude of lakes and pools on the bigger islands. The black-throated divers feed and breed on the larger waters, for example the Holy Lake behind the monastery, whereas the red-throats generally breed on the fringes of small pools and continually fly to and from the sea to feed. Walking in among the boggy birch woods, more of Solovetski's breeding birds are apparent—bramblings, willow tits, redwings and an occasional little bunting, in willow or birch scrub, sometimes easily located by its brief, cheerful but tuneless song.

Outside the breeding season the icy waters are the winter resort of up to 1,000 rare Steller's eiders, migrants from nesting grounds further east along the frozen Siberian coasts. The islands also support three or four resident pairs of the equally rare white-tailed sea eagle. Harbour porpoises and common seals thrive in the waters of the White Sea and Onega Bay and occasionally a fin whale or a minke finds its way in from the Barents Sea. However, the most important large mammal for which the area is famous is the little white whale of the Arctic, the beluga—the 'sea canary'. Although it is always described as 'little' in the context of the family of great whales, an adult can still weigh up to 2 tons. Belukhas, to give them their Russian name, are social animals which live in large groups, sometimes in the thousands and there is a strong population that lives permanently in these food-rich waters, but which has been exploited for countless centuries. Any sea crossing to the islands virtually guarantees the prospect of seeing numbers of them. One after another, their milky white backs arch low over the slate-grey surface of the sea. See them in the early morning light against the rising sun and they are a delicate, almost ethereal pink. You realize slowly just how many there are, away into the distance. They have been hunted in the White Sea since time immemorial—a petroglyph from the area dates back many thousands of years showing a shaman communicating with the white whales. Killing continued up to recent times until it was banned by the Russian Government in 1999.

In summer the whales come close inshore off Solovetski, spending time in the shallows off the aptly named Beluga Cape (Beluzi), where their close proximity can give unrivalled views. Here, each summer, the adults scour their yellowing outer skin on the sea bed to reveal a fresh, pristine whiteness. On the open sea the whales rise and fall gently like the ghosts of Solovetski itself, pale and spectral: the devout spirits of those who lived and built and fought and suffered in this forsaken corner of Russia. The White Sea never freezes over completely, so even in the dark days of winter the belugas do not migrate elsewhere but live on the fringes of the ice, moving with the shifting channels and leads that the tides and winds create. In normal winters wide ice-free margins exist around the islands and along the mainland coasts. Thus the belugas are here at all seasons—spirits of the tortured soul of Solovetski.

FIGURE 19 Beluga.

© Lonely Planet Images/Alamy.

For Solovetski, with its past conflicts hidden behind its natural tranquillity and solemn beauty, endured one even more dreadful chapter in the twentieth century. In 1923, only months after a devastating fire had swept the monastery it became the first, and one of the most infamous, of Stalin's gulags—the vanguard of the Gulag Archipelago—and lasted as such for 20 years. A dark heavy atmosphere overhangs the islands and every modern visitor and pilgrim visiting the place is starkly aware of the history; its very existence was so recent—in the lifetime of many of us. The inescapability of that realization underlies every visit. I have found it impossible to stop visualizing and imagining, at every turn, the horrors of deprivation, torture, and extermination that took place there. The legendary stories of the unthinkable cruelty of the descent into human depravity, driven by the embodiment of calculated evil, are gnawingly indelible. At a peak there were more than 50,000 doomed souls incarcerated here. The cruelties, even apart from the permanent borderline of starvation, were almost unimaginable. The logging gangs were reckoned to offer the worst fate—a short route to hell. Elsewhere, in winter men were forced to walk 500 yards (450 m) to a bath house, naked, at 20° below or sent 2–3 (3–5 km) miles in negligible clothing; they dug the White Sea–Baltic canal by hand in a mere 20 months when over 10,000 died; in summer some suffered the torture of 'death by mosquito', tied naked to a tree or post in boggy areas, to be covered through the night by a blanket of biting mosquitoes, driving men insane or killing them outright. The most minor offence—frequently imagined—could result in hideous punishment.

Such are examples of the endless invention of barbarity devised to amuse the jailers. I walk the woods, pass around the Holy Lake or peruse the magnificent buildings inside the Kremlin and am haunted ceaselessly by the thoughts of what the prisoners endured in each place before their brief tortured lives were ended, many through frozen exhaustion, others summarily shot for no reason and some who agonisingly ended their own lives or mutilated themselves to avoid further unendurable cruelty. In the Kremlin buildings, visitors pass in muted silence through a museum of memories of those dreadful times. The stones of the Kremlin itself still seem to speak of the barbarities to which they were witness; can the memory of evil be erased through the passage of time or will it for ever be an indelible scar, outweighing all the centuries of religious piety?

On islands with such dark memories one looks for lighter thoughts. The monks returned to Solovetski in 1990 and much restoration of the buildings is ongoing. In 1992, the islands were justly declared a UNESCO World Heritage Site, one of the country's architectural jewels. The village alongside the monastery pulses very quietly with the business of everyday life and, in the sun-kissed months of spring and summer, the islands shimmer with the freshness of vegetation, the brightness of wild flowers and the songs of birds.

Solovetski is an extraordinary place, in many ways a microcosm of the history of Russia itself, with milestones in the islands' story marking parallel moments in the wider context of this huge country. They are islands of raw beauty, little known and little visited, with vibrant wildlife, dark history, and rich cultural complexity, but cursed with the haunted shadows of recent times.

St Peter and St Paul Rocks

These isolated rocks, far out in the Atlantic, are the only islands covered in this book on which I have not spent time, nor ever succeeded in landing on. The nearest I have been was standing by the rail and viewing them for two or three hours from a ship lying a hundred yards offshore. St Peter and St Paul Rocks, despite their location and their tiny size, are remarkably interesting in several respects. They are truly remote from other land and, low in the water in mid Atlantic, give the impression of barren sea-washed skerries under continual inundation. However, the facts are a little more interesting than that image suggests. The Rocks help to illustrate the amazing ability of some disparate forms of wildlife to exist and adapt in the most unpromising of environments.

The rocks are situated 60 miles (95 km) north of the equator, 590 miles (950 km) north-east of the Brazilian coast and a little further than that from the coast of West Africa. The nearest point of land is the beautiful island of Fernando de Noronha (Chapter 8), 390 miles (630 km) to the south-west. The St Peter and St Paul Rocks comprise about a dozen minuscule rocky reefs, compressed within an area of sea only 350 yards by 200 yards (320 m × 180 m) and in total amounting to no more than 4 acres (1.6 hectares) of emergent land—a pinprick

FIGURE 20 The Rocks.
© The author.

in the vastness of the ocean. The highest point of land on the appropriately named South-west Rock is a dizzying 60 ft (18 m) above the sea. Their origins once again lie in the violence of the Mid-Atlantic Ridge, the earth's longest sub-oceanic mountain chain, stretching from Iceland in the far north all the way to the southern ocean. Their submarine base is at an enormous depth, some 2¼ miles (3.5 km) below the surface of the sea in the foothills of the mountains but, despite being the direct result of ancient volcanic action, the islands themselves are not composed of volcanic rocks, but of peridotite, which is the dominant rock in the upper part of the earth's mantle. This is the only place on earth where the mantle is exposed above the sea surface. Thus it is demonstrated that the islets were blasted up from the earth's mantle by the force of volcanic action below and appear here as the eroded peaks of the submarine mountain chain.

Unsurprisingly there is virtually no vegetation on the rocks, which are continually washed by salt spray and frequently inundated.

Only on the largest islet, South-west Rock, are there a few grasses and mosses. Otherwise the sea-washed rock surfaces are barren, save for algae. However, they are not devoid of animal life but are inhabited by armies of large marine crabs (*Graspus adscensionis*). These crabs closely resemble the abundant sally lightfoot crab of the eastern coasts of the Americas and were not separated from them as a distinct species until recent decades. They succeed in living in these turbulent waters, feeding on algae and dead sea creatures. Charles Darwin called here in February 1832 during the *Beagle* expedition and his diligent searches on land recorded a feather louse, avian ticks, a woodlouse, some small spiders, and one beetle, as well as the armies of crabs. But, if terrestrial life is sparse, the submarine communities are certainly not, for they are recognized as being particularly rich and are the subject of continuing research (see below). In 1998 the Brazilian government designated the area as a marine protected zone in recognition of the richness of the marine ecosystem around the islands.

Seen from just offshore, under a midday sun, the islets lie low and crouched on the starboard bow. Between the ship and the shores were two or three bright yellow buoys. On each one was a brown booby. Unlike all other members of its clan which are predominantly white, this booby is a dark chocolate brown bird with sharply defined white belly, bright canary feet and a heavy dagger-shaped yellow bill. The perched birds swayed with the movement of the buoys, with shoulders hunched, appearing to be resting their 'heads on their chins'. They have bright penetrating eyes and a somewhat disconcerting all-seeing look. This species is less pelagic than most others of its kind and is more regularly to be found in inshore waters; if so, why are the birds far out at sea here on the Rocks? The answer is that, improbable though it may be on these salt-sprayed rocks, a number of them actually breed here. So too, do a few brown noddies and black noddies, dark relatives of the large family of terns, and there were certainly small numbers of them flying round over the rocks. This also poses an interesting problem. When Charles Darwin called at the islands in 1832 he claimed that he saw a vast multitude of sea-fowl around the Rocks. This interesting observation was echoed 47 years later by H. N. Moseley when working as one of the naturalists on the famous *Challenger* expedition from 1872 to 1876. The expedition principally aimed to verify some of Darwin's assertions about evolution but, more importantly, in effect invented the science of oceanography. When the ship called at St Peter and St Paul Rocks, Moseley similarly claimed that there were birds flying around the Rocks in thousands. Even if this assertion was subject to a degree of embellishment on rocks as small as these, the situation today is very different. Certainly, although the same species are still present today, the numbers are very modest and there has clearly been a substantial decrease in the last 150 years. When Smith (1974) surveyed the islands, he found the seabirds at many different stages of nesting which indicates clearly that here they breed aseasonally with young appearing at any month. It is not likely that the large numbers of crabs are responsible for the decline of seabirds, although without doubt they take some of the eggs when opportunity is presented. The cause of decline is probably attributable to human disturbance for, ridiculous though it may seem, there is a permanent human presence here on these tiny rocky reefs.

On an April day when we anchored there, a group of three fishermen were ashore, sitting on the rocks with their tuna boat tied up to one of the buoys, providing another resting place for a couple more boobies. Needless to say, the fishermen are not resident, but the area is an important tuna fishing ground, with boats regularly coming out from the mainland. More permanently, the Brazilian navy maintains a lighthouse on the crest of South-west Islet, originally built in 1930 and replaced in 1995, where there is also a building housing a small marine scientific station. This facility is permanently manned by relays of four researchers who work a two-week shift and are

FIGURE 21 Brown booby.

© The author.

then relieved by the navy. There is of course, no fresh water on the islands, except what may fall as rain.

Life in the scientific station must be accompanied by degrees of apprehension, bearing in mind the seeming vulnerability of the site and the fact that the Rocks lie in the irregular path of tropical storms and above an unstable tectonic plate. Any unease was clearly not helped in June 2006 when the islands were hit by a powerful earthquake, followed by a tidal surge, flooding the research station, destroying most of it and causing the men to take shelter in the lighthouse. A nearby fishing vessel rescued the four scientists and a new well-equipped and earthquake-resistant station has since been built on the same base for the research work to continue.

20

Tuamotu Archipelago

Through these accounts of twenty islands and archipelagos scattered across the five oceans, we gain an insight into the immense variety and individuality of life that each island and archipelago supports. Most boast classic examples of endemism—either species or subspecies which have evolved there and there alone—thereby vastly enriching the variety of life on the planet. The Canary Archipelago, for example, boasts some 520 endemic plants; the Galapagos Islands are world famous for the evolution of distinct varieties of species from island to island. The dodo existed nowhere except Mauritius, the unique Azorean noctule bat hunts nowhere except over the dry woodlands of the Azores, while most oceanic islands evolved their own species of flightless rail. What remarkably successful pioneers that family has produced! Over the millennia the fauna and flora on remote islands prospered and flourished there in a stable state, answering only to the forces of nature. That is, of course, until the arrival of man over the past 500 years or so, which had a devastating effect on each fragile Eden that he discovered. Since then natural systems have been disrupted or destroyed, species have been lost and in some cases islands have then been abandoned as being no longer useful.

Over recent decades the wheel has turned and by the beginning of the twenty-first century the consciousness of what has been lost, the importance of island biodiversity and the need to take positive steps to recover some of the losses has become an increasing crusade; various examples are to be found in preceding accounts in this book. The bottom line of any initiative aimed at retrieving threatened species and restoring populations, is the need for accurate information on numbers and status. In terms of birds alone, there are over 10,000 species in the world (the number increases as scientists continue to redefine and separate existing species). Of these, 1,240 are now regarded as threatened with extinction (i.e. 'critically endangered', 'endangered', or 'vulnerable'), with another 838[1] classed as 'near threatened'. In addition at least 130 species have become extinct since AD 1500. A disproportionate number of these worldwide totals refer to birds on oceanic islands and a high percentage of those are in the Pacific—although of course this is where the majority of oceanic islands are to be found. On many of these isolated island groups there is a great need for up to date information about the wildlife there.

Thus it was, a few years ago, that I spent time in the Tuamotu Archipelago in the far Pacific, with three colleagues, travelling from atoll to atoll, trying to establish the baseline status of several bird species about which desperately little was known. It was important to ascertain the existence or otherwise of five particular species which are, or were, emblematic of the archipelago: the colourful Tuamotu kingfisher, delicate Tuamotu ground dove, the tiny blue lorikeet, the impressive imperial pigeon, and, especially, the strange and almost mythical Tuamotu sandpiper. Did they still exist? If so where and how many and what should be done to protect them?

The Tuamotu islands are everyman's vision of a tropical South Seas idyll. Here, on the sun drenched, palm shaded atolls, the islanders live in intimate relationship with the wildlife around them. They thrive on the produce of the sea, harvest modest numbers of seabirds for the pot, exploit the crucial bounty of introduced coconut palms and cherish the importance of the gift that nature has bestowed on them in the form of beds of valuable natural pearls. They live by the sea, on the sea, and are essentially of the sea. Every aspect of their lives is tied to the different moods of the ceaseless ocean.

There are seventy-eight individual atolls scattered over 900 miles (1,500 km) of sea, making the Tuamotus the largest archipelago in the world, although its remoteness means that the name may be far less familiar than many other Pacific groups. Every atoll is essentially a huge ring of coral, each formed in far distant eras on the flanks of a volcano, long since eroded and lost below the waves, now leaving the atolls palm-decked, with crystal sands, cerulean skies, a shimmering clear lagoon at their hearts and the perpetual thunder of surf on the outer shores. An atoll's encircling coral ring comprises a necklace of small islands—motus—vegetated mainly with coconut palms, and each separated from the next by knee-deep channels—hoas. Some individual atolls are very large. Rangiroa, at the northern end of the chain with a circumference of 175 miles (280 km) is the second largest atoll in the world after Kwajalein in the Marshall Islands. The Archipelago lies 250 miles (400 km) east of Tahiti and runs from north-west to south-east, almost wholly within the Tropic of Capricorn. A handful of the atolls are thinly populated although the majority—including the smaller ones—are not. The Tuamotus have always had a bad name among sailors for the serious dangers they pose to navigation and they were long ago nicknamed 'The Dangerous Isles'. In our time they have two particular claims to fame. In 1947 Thor Heyerdahl sailed his raft Kon Tiki across the ocean from South America and eventually made a successful landfall at the small community on the eastern atoll of Raroia. Less happily, from the 1960s until the 1990s the French used the twin islands of Mururoa and Fangataufa at the southern end of the archipelago as their nuclear testing site, much to the eternal offence of the Paumotans. Both islands still remain uninhabitable.

Like other oceanic islands, the Tuamotus have a distinctive flora and fauna, with many endemic species. The picture was undoubtedly more impressive in the past, for the numbers of some species are much reduced, while others have inexorably been lost since the arrival of man, first Polynesian man and latterly European man accompanied by introduced rodents and other mammals in the eighteenth century. Vast distances from the nearest continental mainlands mean that there are no native mammals here, but there are many interesting plants and threatened bird species about which insufficient information exists.

Polynesians are known throughout the world as magnificent seamen and navigators and the Paumotans are no exception. Even if outrigger canoes on the Tuamotus have given way to tiny wooden craft with outboard motors, the men still make eye-watering journeys across the ocean from one atoll to another which is often invisibly beyond the horizon. They travel with nothing aboard other than a spare fuel can and a 'shark pole' (the purpose of which it seems impossible to ascertain). On these—and hundreds of other remote Pacific atolls—the people live not only on the edge of the sea, but in many ways as an evidently integral part of that environment. As on so many other islands, they live in harmony with nature, their lives dependent on the daily produce of the land and sea. Certainly on the sparsely inhabited islands, away from access to other supplies, fish is a staple, hunted and eaten fresh every day in one form or another, cooked or raw. On such atolls, coconuts are harvested constantly (and can be distilled) and the only crops grown are often tiny patches of sweet potatoes and onions. Occasional forays are made across the lagoons to colonies of boobies or frigate to harvest youngsters (or adults) to vary the monotony of a fish diet. Islanders such as these have no requirement for material goods and live the simplest of lives, happy just living.

Toau is a classic example of this pattern of life. The community on this atoll was small during our visit, five men and four women in an extended family, living in three small houses among the palms behind the landing beach, close to the side of the deep-water passage which on all atolls connects the lagoon with the open sea. In the early years of the current century the community is scarcely larger, simply expanded with the addition of one or two members of a younger generation. These fine people are classic Polynesians, superb physical specimens, black haired and dark skinned, the men bare to the waist with broad shoulders, barrel chests, powerful arms and legs. The women, similarly, are handsome, powerfully built Amazons, with huge smiles to match. On our visit to Toau, it was salutary to realize that they lived permanently and happily on one small motu, in this remote place with no means of contact with the outside world. A copra boat travelling from island to island calls occasionally, from which basic food items such as rice and chickens could be purchased and which in return would buy their catch of fish or harvest of copra. But such is the nature and pace of life in the Tuamotus that sometimes the boat comes, sometimes it doesn't.

The Tuamotus are the heartland of the legendary pearl fishermen and are famed particularly for the fabulous black pearls. These pearls, growing in the pollution-free waters of the lagoons, have been harvested for generations by native pearl fishers. They could descend to remarkable depths in the lagoon, in the search for the *Pinctada* oysters in which the pearls develop. The families on Toau atoll are immediate descendants of the last generation of 'professional' divers. The two women habitually dived to 80 ft (25 m). Their father was one of the most famous Tuamotu pearl divers and claimed a depth of 130 ft (40 m), staying under for over three minutes. Sadly most of this unique tradition is history because, since the 1960s, the artificial culture of pearls has been developed in the lagoons and is now the principal means by which they are commercially cultivated. Cultured pearls remain a valuable product from the Tuamotus with the majority being sold at auctions in Tahiti twice a year, raising upwards of $10,000,000.

There is an air of genuine remoteness wandering islands such as these: a tingling excitement that one really is out of contact and out of reach of the rest of the world. The tiny community on Toau, like so many others, lives on one small motu, although the atoll itself has a circumference of over 20 miles (32 km), with roughly the same number of individual motus. Thus there are great stretches of ground that are very rarely visited and many that have barely felt the presence of man's feet. On all the islands the interior of the motus is dominated by coconut palms and the ground is strewn with a dense carpet of generations of fallen nuts. Although the palms over-top all other growth, there are still native trees, among them *messersmidia* and delicate *pandanus* are the most numerous. Everywhere there is a rich understorey of shrubs including *morindas* and bright, flowering *scaevolas* in which the boobies nest at the head of the beaches. Among the tangled vegetation under the coconut palms an island endemic, the Tuamotu reed warbler, is common. One can spend days crossing a succession of hoas and walking miles along the coral beaches on the ocean side of the motus and returning on the calmer sands of the lagoon edge. The hoas, open to the ocean on one side and the lagoon on the other, are perhaps 100–200 yards (90–180 m) of knee-deep water and sound daunting although the shallow waters are generally safe to cross provided that one is properly shod. They are home to myriad fish and other sea life. Large rays, baby sharks, and eels scuttle harmlessly away, but the hoas are also home to a breathtaking catalogue of deeply unpleasant animals whose poisons are seriously dangerous—stonefish, cone shells, sea anemones, and sea snakes.

The vulnerability of tiny communities in isolated places was well demonstrated on Toau when the island was caught in the eye of a violent cyclone with winds of terrifying force. The whole island was effectively wrecked. On the occupied motu thousands of trees snapped and others were blown down and disappeared to sea, the dwelling huts vanished completely and everything freestanding was gone. The islanders survived only by smashing a hole in the bottom of the concrete rain-water tank and taking refuge inside. Here they stayed for several days, until relief came and the storm had passed. Life in paradise is not always an idyll. With no communication to the outside world a great fear is always serious injury. Most of the men bore scars from shark attack but they were just as wary of barracudas, with their fearsome armouries of teeth, which are sometimes caught in their fish traps.

One of the understandable characteristics of island communities is that they invariably have a sound knowledge of the wild plants and creatures amongst which their lives are lived, and in many cases, on which they are heavily dependent. The Tuamotan island of Apataki supports a small community of a couple of hundred souls and is a case in point. Shown pictures of appropriate bird species, islanders like these usually give a fairly accurate indication of the presence and status of different ones. If they claim no knowledge of a particular species, it is fairly good evidence that it is unlikely to occur there. A near neighbour to Apataki is the little island of Niau. In 1989 islanders verified the fact that the emblematic bird on that island, the Tuamotu kingfisher, existed in good numbers. Remarkably, although Niau is in easy view from several other atolls, the kingfisher now occurs there and nowhere else in the world. It is a magnificent bird with black cap, white body, and brilliant turquoise wings and feeds on skinks, geckos, and arthropods. Now, in the twenty-first century, the kingfisher population living in the coconut plantations on Niau comprises only about 125 birds, but is stable, although, with its tiny population, it remains one of the species still classified as critically endangered. In much the same vein we discovered a population of c.300 individuals of the rare and exquisite 'vini', the blue lorikeet, on Tiamanu Motu on Apataki. It has subsequently been found on several other atolls in the north of the archipelago giving a total of some 3,000 birds for the Tuamotus. On Apataki it is valued not only as a wild bird, but also as a pet and several are kept as pets in the village. It is one of the wider family of fifty-five loris and lorikeets, all of which are small, brightly-coloured parrots found scattered individually throughout Pacific islands. They feed on soft fruit and nectar which they extract from the plants with the use of specially designed brush-tipped tongues. The 'vini' has been lost from other sites in the Society Islands although good numbers still occur on Aitutaki in the distant Cook Islands where it was evidently introduced long ago.

The lonely beaches, hoas, and miles of lagoon sides of the Tuamotus are home to a scattering of interesting birds, always thinly distributed and usually only found singly. The reef herons sitting, hunched, waiting for the waters to ebb in tidal lagoons come in two plumages, either a white morph or the much darker grey form although, to confuse the picture further, some individuals are intermediate between the two extremes. They are solitary birds, hunting by stalking and picking up a variety of shoreline creatures with a preference for small shore crabs and diminutive coral fish. Pacific golden plovers, long distance travellers from the Arctic, patrol the open beaches but especially favour muddy pools among the palms on the lagoon side of the motus. Small groups of black noddies stand resting on the coral strands and wandering tatlers, grey-bodied, yellow-legged shorebirds from nesting grounds along riversides in Alaska are frequent on these islands during the northern winter and forage among the coral litter on the shoreline. Occasional bristle-thighed curlews, another breeding species from distant Alaska, also occur on these shores and live solitary, individual lives on the beaches for the months they are here. This handsome curlew—a real birdwatcher's bird—with bright rusty tail, is unique among all shore birds in that it moults all its flight feathers at the same time, with the result that for a few weeks on these wintering grounds it is completely flightless, while waiting for the new feathers to grow. It is for this reason that the birds make the huge ocean crossing to remote and lonely Pacific atolls such as these, where they are safe from land predators. It is quite likely that it will eventually be shown that these curlews make the long flight of over 3,000 miles (5,000 km) across the Pacific from Alaska, non stop, as has been shown with other wading birds from the same breeding areas flying to wintering grounds south of the tropics.

Overhead on the atolls, especially in the evenings, straggling lines of red-footed boobies return to their nests from feeding excursions at sea, together with individual brown boobies, great frigatebirds, and noddies. Most seductive of all the seabirds on these and many other tropical island groups are the exquisite fairy terns (now less appealingly called white terns). These delicate, pure white birds 'nest' in the most ridiculous places, for they build no nest at all, but lay a single egg in crooks on bare branches, on coral rocks, and similar completely exposed sites. They fly to and fro, moth-like, with the bright light shining through their delicate wings.

However, amongst these various birds there is no sign on any of the islands of the elusive Tuamotu sandpiper, although in 1989 some old islanders had 'heard tell of the strange elusive titi' (Paumotan name for the Tuamotu sandpiper). On Fakarava, two old fishermen talked birds to us. They knew nothing about the two pigeons or the lorikeet, but were positive about the 'titi'. Despite the severe cyclone in the mid 1980s they believed that this mysterious bird still survived on some of the islets in the middle of the huge lagoon on Fakarava— islets with magical, evocative names—Hapiapia, Panu Panu, Kiria, Koaka Ruga. Sadly for us there was no fuel (or local inclination?) for any visit there.

Fakarava is a very large atoll. Across the world, every island has its own atmosphere: in the Tuamotus Apataki pulses with life, Toau is overwhelmingly friendly and as laid back as it is possible to find. Fakarava exudes a grim eerie chill. It is the most disconcerting island I have ever been on. My reflections on it precisely echo Robert Louis Stevenson's account of his visit there a hundred years earlier. There is an overwhelming sense of foreboding and almost calculated coldness from the moment of arrival. I would not happily return. Our arrival those years past was unexpected but, far from the usual instantly curious crowd, nobody appeared from any of the huts to enquire or even just to look. We knew there was a community there but it might as well have been a ghost island. A tangible dismissivenesss pervaded the whole place. We learned later that an incident of almost medieval proportions had occurred a few years earlier which, at the time, reverberated around the world. In a dispute originating in religious fanaticism, men from a nearby island visited Fakarava and the resulting conflict involved torture, murder, ritual burnings, and even the dragging of one victim with a hook through his tongue. The event seemed to epitomise the dark immanence of the island.

The Paumotans on this beautiful archipelago live a fairly high risk existence. Although they necessarily live in daily interdependence with the sea, it is often also their nemesis. When we were on one of the larger islands, the community of 360 had lost eight men at sea in the previous eight weeks. That island, Makemo, boasts the highest point of land on all the islands, one small coral ridge 17 ft (5 m) high. If the worst predictions of climate change occur, their futures will be beyond question. Tatlers and crested terns inhabit these ridges but many of the original endemic birds on this island as on others, have evidently succumbed to the presence of rats, cats, and pigs while others like the lorikeet and the kingfisher are severely restricted and their fate hangs perilously in the balance.

It is the inevitable ending to all good stories about searches in remote places for some rare creature or another that the successful outcome occurs only at the very last moment. On the last morning of our time on the Tuamotus, we drifted lazily into the lagoon of a beautiful, deserted atoll and eased our way towards a sandy beach under a cloud of frigate birds and boobies from their breeding colony on one of the motus. As our boat glided in, four tiny birds flew out and circled us, calling 'titi, titi, titi'. The Tuamotu sandpiper survives at least here, but almost certainly, as we learned from the boatman, on some of the other uninhabited small atolls nearby as well. It is a bizarre little bird, a misfit among shorebirds. It has grey-brown mottled plumage with dark yellowish legs and a small pointed bill; it exhibits no fear of man because on this atoll it has no experience of him. It feeds among the coral litter at the top of the beach but also walks around on the twigs and branches of the *pandanus* bushes picking off ants and other small insects. No pigs, no rats, no cats, and no regular human presence endanger it on the few islands where it survives. However, the existence of the sandpiper on this atoll leaked soon after we found it and since then cruise ship birdwatchers have visited every year. Man's presence ever expands. How long will this small Eden exist? The Tuamotus are now well on the tourist trail anyway. Even Fakarava boasts tourist accommodation. But would I willingly visit that particular island again? No, I think not.

EPILOGUE

What is the future for islands such as those described in this book?

It is of course impossible to predict, although there are some strong pointers. One of the certainties is that life on many islands is changing and will continue to change. The twenty-first century boom in cruise travel is a phenomenon which makes virtually anywhere accessible and the more remote and unvisited a place is, the more attractive is its appeal. Large cruise liners widen their horizons all the time while the smaller expedition ships, provided with highly manoeuvrable zodiacs, can reach beaches and shorelines around the world that larger ships can only dream of.

Since I first visited some of the islands featured in this book I have seen rapid developments taking place. Cruise liners are more regular now, even in the White Sea in northern Russia, where more tourists are visiting the Solovetski islands each year. Around the San Blas Islands in the Caribbean increased numbers of visitors will undoubtedly change the islanders' priorities and perceptions. Even on remote Tristan da Cunha provision is now made for residential visitors as it is on an increasing number of the Tuamotus. Graciosa has already lost its innocence, as mentioned in the account of Chinijo Archipelago, although the recent binding conservation designations on all the islands of the group should ensure that no further developments occur and its rocky neighbours remain as wild as they are now. Wisely, landing on ice-bound Halfmoon Island has now been stopped to protect the historic human and animal remains. Great Skellig, oceanic though it is, is another fragile environment which has necessitated a strict restriction on the numbers of visitors, whilst Fernando de Noronha has already limited the number of tourists it will accommodate at any one time. The long-term problem there, however, may be more to do with its rising human population.

The days of ruthless exploitation of island wildlife have hopefully passed into an uncomfortable chapter of history, but many serious problems still remain, such as uncontrolled overfishing worldwide and its knock-on effect on marine populations of birds and mammals. Increased eco-tourism on fragile island sites has to be carefully managed; the solutions on Great Skellig and Halfmoon Island being two examples. The visitor regime on South Georgia is another model designed to allow visitors to experience the magnificent wildlife onshore while ensuring its protection. Behind these issues lurks the ever present risk of environmental disasters such as that caused by the unprecedented shipwreck and resulting disastrous oiling of seabirds on Nightingale Island in mid ocean, 20 miles (32 km) from Tristan da Cunha, in 2011. Much commendable, but expensive, pioneering work is being undertaken to restore habitats, vegetation communities and populations of birds and other species on many islands. We need to ensure that these initiatives are not jeopardized by eco-tourism or other man-made pressures. Certainly the progress made, for example, on South Georgia and Ascension (and especially Henderson Island, not covered in this book) is encouraging but much remains to be done on a host of other islands.

What then of the human populations on remote island stations? Changes across even the most isolated of oceanic islands are happening fast as technology and increased mobility bring them into ever closer contact with the modern world. This fact begs one slightly uncomfortable question. Do the changes which we might assume improve the quality of life actually bring more contentment to the groups of people living there? Life in the slow lane has undoubtedly accelerated but are the three or four families of fishermen who alone populated Graciosa in the 1970s, happier and more fulfilled now than they were before development overtook their island? On Tristan da Cunha the same seven family names occur as they did before the enforced evacuation when the volcano unexpectedly exploded in 1961, but their life style now is dramatically different. Every cottage has electricity, the hospital is modernized, visitor accommodation has been built, and a policeman appointed. All these developments are clearly beneficial but reliance on a cash economy, increasing numbers of motor vehicles (with almost nowhere to drive) and the introduction of income tax make me wonder if any of the older inhabitants hanker for the days of slower life, bullock carts as transport, oil lamps for lighting, and a community where everybody helped with whatever tasks were needed and payment was not part of the equation?

The urge to return to island living is well exemplified by the Chagos islanders who, after decades of forced exile, continue with their demand to be allowed to return to their traditional homes on the Indian Ocean archipelago. Three years after resettlement following the volcano eruption, several families of Tristan islanders went to live in the UK but most soon realized that they missed the island life too much and returned home once more.

It is easy for us to see isolated islands with a rosy glow and envisage lives of contentment, if not ease, enjoyed by their inhabitants as they maintain their traditional and unique ways. What is probably inevitable, however, is that, as the world shrinks, individuality will diminish and distant communities, increasingly drawn into the wider world, will slowly become homogenized and lose much of their distinctive features.

APPENDIX
Scientific names of species

Scientific names of species mentioned in the text

BIRDS

Penguins—*Sphenisciformes*
King penguin	*Aptenodytes patagonicus*
Northern rockhopper penguin	*Eudyptes moseleyi*

Divers—*Gaviformes*
Black-throated diver	*Gavia arctica*
Red-throated diver	*Gavia stellata*

Tubenoses—*Procellariiformes*
Tristan albatross	*Diomedea dabbenena*
Yellow-nosed albatross	*Thalassarche chlororhynchos*
Great shearwater	*Puffinus gravis*
Manx shearwater	*Puffinus puffinus*
Cory's shearwater	*Calonectris diomedea*
Little (Macaronesian) shearwater	*Puffinus baroli*
Bulwer's petrel	*Bulweria bulwerii*
Spectacled petrel	*Procellaria conspicillata*
White-chinned petrel	*Procellaria cinerea*
Southern giant petrel	*Macronectes giganteus*
Northern giant petrel	*Macronectes halli*
British storm petrel	*Hydrobates pelagicus*
Madeiran storm-petrel	*Oceanodroma castro*
Leach's petrel	*Oceanodroma leucorrhoa*
White-faced storm-petrel	*Pelagodroma marina*
Wilson's storm-petrel	*Oceanites oceanicus*
Fulmar	*Fulmarus glacialis*
Diving petrels	*Pelecanoides spp*

Gannets, etc.—*Pelicaniiformes*

Red-billed tropic bird	*Phaethon aethereus*
Yellow-billed tropic bird	*Phaethon lepturus*
Northern gannet	*Morus bassanus*
Red-footed booby	*Sula sula*
Brown booby	*Sula leucogaster*
Masked booby	*Sula dactylatra*
Great cormorant	*Phalacrocorax carbo*
Shag	*Phalacrocorax aristotelis*
Brown pelican	*Pelecanus occidentalis*
Magnificent frigatebird	*Fregata magnificens*
Ascension Island frigatebird	*Fregata aquila*

Herons and allies—*Ciconiiformes*

Dimorphic egret	*Egretta dimorpha*
Little egret	*Egretta garzetta*
Cattle egret	*Bubulcus ibis*
Ascension Island night heron	*Nycticorax olsoni*
Pacific reef heron	*Egretta sacra*

Wildfowl—*Anseriformes*

Lesser snow goose	*Chen caerulescens*
Common eider	*Somateria mollissima*
Steller's eider	*Polysticta stelleri*
American wigeon	*Anas americana*
South Georgia pintail	*Anas g. georgica*

Birds of prey—*Falconiformes*

White-tailed sea eagle	*Haliaeetus albicilla*
(American) Black vulture	*Coragyps atratus*
Common buzzard	*Buteo buteo*
Golden eagle	*Aquila chrysaetos*
Gyr falcon	*Falco rusticolus*
Peregrine	*Falco peregrinus*
Eleonora's falcon	*Falco eleonorae*
Hobby	*Falco subbuteo*
Sooty falcon	*Falco concolor*
Mauritius kestrel	*Falco punctatus*

Pheasants, etc.—*Galliformes*
    Ptarmigan   *Lagopus mutus*

Rails—*Gruiformes*
    Black francolin   *Francolinus francolinus*
    Tristan moorhen   *Gallinula nesiotis*
    Gough Island moorhen   *Gallinula comeri*
    Inaccessible Island rail   *Atlantisia rogersi*
    Red rail   *Aphanapteryx bonasia*
    Guam rail   *Gallirallus owstoni*

Gulls, Auks, Waders—*Charadriiformes*
    Oystercatcher   *Haematopus ostralegus*
    Canarian black oystercatcher   *Haematopus meadewaldoi*
    Ringed plover   *Charadrius hiaticula*
    Pacific golden plover   *Pluvialis fulva*
    Grey plover   *Pluvialis squatarola*
    Semi-palmated plover   *Charadrius semipalmatus*
    Kentish plover   *Charadrius alexandrinus*
    Kildeer   *Charadrius vociferous*
    Lesser yellow-legs   *Tringa flavipes*
    Purple sandpiper   *Calidris maritima*
    Knot   *Calidris canutus*
    Dunlin   *Calidris alpina*
    Pectoral sandpiper   *Calidris melanotos*
    Snipe   *Gallinago gallinago*
    Bristle-thighed curlew   *Numenius tahitiensis*
    Whimbrel   *Numenius phaeopus*
    Willet   *Tringa semipalmata*
    Turnstone   *Arenaria interpres*
    Tuamotu sandpiper   *Prosobonia cancellata*
    Great skua   *Stercorarius skua*
    Brown skua   *Catharacta antarctica*
    Long-tailed skua   *Stercorarius longicaudus*
    Arctic skua   *Stercorarius parasiticus*
    Yellow-legged gull   *Larus michahellis*
    Great black-backed gull   *Larus marinus*
    Lesser black-backed gull   *Larus fuscus*
    Glaucous gull   *Larus hyperboreus*

Ring-billed gull	*Larus delawarensis*
Laughing gull	*Larus atricilla*
Kelp gull	*Larus dominicanus*
Common gull	*Larus canus*
Ivory gull	*Pagophila eburnea*
Kittiwake	*Rissa tridactyla*
Arctic tern	*Sterna paradisaea*
White tern ('fairy tern')	*Gygis alba*
Black noddy	*Anous minutus*
Brown noddy	*Anous stolidus*
Greater crested tern	*Sterna bergii*
Bridled tern	*Onychoprion anaethetus*
Sooty tern	*Sterna fuscata*
Great auk	*Pinguinus impennis*
Razorbill	*Alca torda*
Guillemot	*Uria aalgae*
Brunnich's guillemot	*Uria lomvia*
Puffin	*Fratecula arctica*
Little auk	*Alle alle*
Black guillemot	*Cepphus grylle*
Snowy sheathbill	*Chionis alba*

Pigeons—*Columbiformes*

Dodo	*Raphus cucullatus*
Rock Dove	*Columba livia*
(Polynesian) Imperial pigeon	*Ducula aurorae*
Tuamotu ground dove	*Gallicolumba erythroptera*
Zebra dove	*Geopelia striata*
Eared dove	*Zenaida auriculata*
Pink pigeon	*Nesoenas mayeri*

Parrots—*Psitaciformes*

Echo parakeet	*Psittacula eques echo*
Blue lorikeet	*Vini peruviana*
Ring-necked parakeet	*Psittacula krameri*

Owls—*Strigiformes*

Snowy owl	*Bubo scandiacus*

Swifts and Hummingbirds—*Apodiformes*

Guam swiftlet	*Collocalia bartschi*

Kingfishers—*Coraciiformes*

Tuamotu kingfisher	*Halycon gamberi*
Micronesian kingfisher	*Todirhampus c. cinnamomina*

Passerines—*Passeriformes*

Meadow pipit	*Anthus pratensis*
Berthelot's pipit	*Anthus bethtelotii*
South Georgia pipit	*Anthus antarcticus*
Grey wagtail	*Motacilla cinerea*
Skylark	*Alauda arvensis*
Wren	*Troglodytes troglodytes*
St Kilda wren	*Troglodytes t. hirtensis*
Northern Wheatear	*Oenanthe oenanthe*
Fuerteventura chat	*Saxicola dacotiae*
Blackbird	*Turdus merula*
Redwing	*Turdus iliacus*
Norohna vireo	*Vireo gracilirostris*
Guam flycatcher	*Myiagra freycineti*
Spotted flycatcher	*Muscicapa striata*
Noronha elaenia	*Elaenia ridleyana*
Willow warbler	*Phylloscopus trochilus*
Tuamotu reed warbler	*Acrocephalus caffer*
Spectacled warbler	*Sylvia conspicillata*
Willow tit	*Parus montanus*
Red-whiskered bulbul	*Pycnonotus jocosus*
Indian myna	*Acridotheres tristis*
Madagascar fody	*Foudia madagascariensis*
Mauritius fody	*Foudia rubra*
Mauritius olive white-eye	*Zosterops chloronothus*
Guam white-eye	*Zosterops conspicillatus*
Red-billed chough	*Pyrrhocorax pyrrhocorax*
Mariana crow	*Corvus kubaryi*
Raven	*Corvus corax*
Starling	*Sturnus vulgaris*
Arctic redpoll	*Carduelis hornemanni*
Canary	*Serinus canaria*
Brambling	*Fringilla montefringilla*
Trumpeter finch	*Rhodopechys githaginea*
Little bunting	*Emberiza pusilla*

Lapland bunting	*Calcarius lapponicus*
Snow bunting	*Plectrophenax nivalis*
Tristan bunting	*Nesospiza acunhae*
Tennessee warbler	*Vermivora peregrina*

MAMMALS

Ungulates—*Artiodactyla*

Reindeer	*Rangifer tarandus*
Musk ox	*Ovibos moschatus*
Soay sheep	*Ovis aries*
Java deer	*Cervus timorensis*
Asian water buffalo	*Bubalus bubalis*

Carnivores—*Carnivora*

Polar bear	*Ursus maritimus*
Arctic fox	*Alopex lagopus*
Wolverine	*Gulo gulo*
American mink	*Neovison vison*

Whales and Dolphins—*Cetacea*

Bowhead	*Balaena mysticetus*
Southern right whale	*Eubalaena australis*
Northern right whale	*Eubalaena glacialis*
Grey whale	*Eschrichtius robustus*
Minke whale	*Balaenoptera acuotrostrata*
Blue whale	*Balaenoptera masculus*
Fin whale	*Balaenoptera physalus*
Humpback	*Megarapta novaeangliae*
Sperm whale	*Physeter macrocephalus*
Beluga	*Delphinapterus leucas*
Long-finned pilot whale	*Globicephala melas*
(Long-snouted) Spinner dolphin	*Stenella longirostris*
Harbour porpoise	*Phocoena phocoena*

Seals—*Pinnipedia*

Ringed seal	*Pusa hispida*
Bearded seal	*Erignathus barbatus*
Harbour Seal (common seal)	*Phoca vitulina*
Leopard seal	*Hydrurga leptonyx*
Southern elephant seal	*Mirounga leonina*

Walrus — *Odobenus rosmarus*
    Antarctic fur seal — *Arctocephalus gazella*

Bats—*Chiroptera*
    Azorean noctule — *Nyctalus azoreum*
    Marianas fruit bat — *Pteropus mariannus*

Lagomorphs—*Lagomorpha*
    Rabbit — *Oryctolagus cuniculus*
    Mountain hare — *Lepus timidus*

Elephants, etc.—*Proboscidea*
    Woolly mammoth — *Mammuthus primigenius*

Rodents—*Rodentia*
    Black rat — *Rattus rattus*
    Brown rat — *Rattus norvegicus*
    Polynesian rat — *Rattus exulans*
    St Kilda house mouse — *Mus musculus muralis*
    St Kilda field mouse — *Apodemus sylvaticus hirtensis*
    Mykines house mouse — *Mus musculus mykinessciensis*
    Collared lemming — *Dicrostonyx torquatus*
    Siberian (Wrangel) lemming — *Lemmus portenkoi*
    Rock cavy — *Kerodon rupestris*

Insectivores—*Insectivora*
    Indian house shrew — *Suncus morinus*

Hyraxes—*Hyracoidea*
    Rock hyrax — *Procavia capensis*

Amphibians
    Cane toad — *Bufo marinus*

Snakes and Lizards—*Squamata*
    Argentine black and white tegu — *Tupinambis merianae*
    Indian wolfsnake — *Lycodon aulicus*
    Brown treesnake — *Boiga irregularis*
    Ornate day gecko — *Phelsuma ornata*
    Telfair's skink — *Leiolopisma telfairii*
    Noronha skink — *Trachylepis atlantica*
    Curious skink — *Carlia fusca*

Turtles and Tortoises—*Testudines*

Giant tortoise (Aldabra)	*Geochelone gigantea*
Giant tortoise—extinct, Mauritius	*Cylindraspis inepta*
Giant tortoise—extinct, Mauritius	*Cylindraspis triserrata*
Green turtle	*Chelonia mydas*
Hawksbill	*Eretmochelys imbricate*

PLANTS

Flowering plants

Marsh marigold	*Caltha palustris*
Pasque flower	*Pulsatilla spp*
Buttercups	*Ranulucus spp*
Glacier buttercup	*Ranunculus glacialis*
Arctic buttercup	*Rananculus hyperboreus*
Arctic poppy	*Papaver spp*
Svalbard poppy	*Papaver dahlianum*
Scurvy grass	*Cochlearia officinalis*
Bladder campion	*Silene vulgaris*
Red campion	*Silene dioica*
Ragged robin	*Lychnis flos-cuculi*
Sea blite	*Suaeda maritime*
Sea orache	*Atriplex halimus*
Glasswort	*Salicornia fruiticosa*
Geraniums	*Geranium spp*
Wood cranesbill	*Geranium sylvaticum*
Meadowsweet	*Filipendula ulmaria*
Pink dryas	*Dryas punctata*
Greater burnet	*Acaena magellanica*
Silverweed	*Potentilla anserina*
Tormentil	*Potentilla erecta*
Tufted saxifrage	*Saxifraga cespitosa*
Bog saxifrage	*Saxifraga hirculus*
Fireweed (Rosebay)	*Epilobium angustifolium*
Angelica	*Angelica sylvestris*
Sorrel	*Rumex spp*
Heather	*Calluna vulgaris*
Cross-leaved heath	*Erica tetralix*
Bilberry	*Vaccinium uliginosum*
Crowberry	*Empetrum nigrum*

Thrift (Sea pink)	*Armeria maritima*
Lousewort	*Pedicularis sylvatica*
Butterwort	*Pinguicula vulgaris*
Dandelions	*Taraxacum spp*
Yellow flag	*Iris pseudacorus*
Heath spotted orchid	*Dactylorchis maculata*
Ile aux Aigrettes orchid	*Oeoniella aphrodite*
Arum lilies	*Zantedeschia spp*
Pink paintbrush	*Castilleja elegans*
Alpine arnica	*Arnica alpina*
Mesembryanthemums	*Mesembryanthemum spp*
Nasturtium	*Nasturtium officinale,*

Non-flowering plants

Tussock grasses	*Spartina arundinacea* and *Parodiochloa flabellate*
Antarctic hairgrass	*Deschampsia Antarctica*
Purple moor grass	*Molinia caerulea*
Ascension Island parsley fern	*Anogramma ascensionis*

Trees

Bois de fer	*Sideroxylon boutoniarum*
Bois de chandelle	*Dracaena concinna*
Mauritius ebony	*Diospyros egrettarum*
False acacia	*Leucaena leucocephala*
(Tristan) 'Island tree'	*Phylica Arborea*
Tangan tangan	*Leucaena leucocephala*

FISH

Barracudas	*Sphyraena spp*
Butterfly fish	*Chaetodontidae spp*
Clownfish	*Amphiprioninae spp*
Damsel fish	*Pomacentridae spp*
Grouper	*Serranidae spp*
Jack fish	*Carangidae spp*
Parrot fish	*Labridae spp*
Sand eel	*Ammodytes marinus*

CRUSTACEANS

Tristan rock lobster	*Jasus tristan*

MOLLUSCS

Gastropods—*Mollusca*
  Giant African land snail   *Achatina fulica*

The bird which occurs on all of the islands in this book is the turn-stone *Arenaria interpres*.

NOTES

Introduction

The English names of some birds have recently been reappraised in order to help avoid confusion as more of us travel abroad and encounter species in the same families as those at home. Thus, for example, the fulmar in British waters is often referred to now as the 'northern fulmar'. However, in this book I have retained the traditional names, as I believe that until the new ones have been fully accepted and are in more general use, a degree of confusion may be avoided by retaining the more familiar names.

A list of the scientific names of species mentioned in the book is given in the appendix.

One species of bird occurs on every one of the islands described in this book; it is named at the end of the appendix.

8 Fernando de Noronha

1 TAMAR – is short for *tartaruga marinha*, the Portuguese term for sea turtles.

13 Mingulay

1 All seabird numbers are estimates from Scottish Natural Heritage surveys carried out on one day in 2008. All species have shown declines of between 12 per cent and 42 per cent since the previous surveys in 2003.
2 Much of the historical detail here leans heavily on Buxton's (1995) book, *Mingulay. An island and its people* and to a lesser extent *The Scottish Islands* by H. Haswell-Smith (2008).

14 South Georgia

1 Much of the historical detail is based on Burton's (2005) booklet *South Georgia*.

17 ILE AUX AIGRETTES

1 The following bird species are extinct on Mauritius: Dodo, Blue Pigeon, Broad-billed Parrot, Grey Parrot, Mascerene Coot, Mauritian Duck, Mauritius Owl, Mascerene Swan, Mauritius Night Heron, Red Rail. Extant endemic species are Black Bulbul, Cuckoo Shrike, Echo Parakeet, Pink Pigeon, Mauritius Kestrel, Mauritius Fody, Grey and Olive White-eyes and Paradise Flycatcher.

20 TUAMOTU ARCHIPELAGO

1 2011 figures. All figures approximate as numbers are constantly adjusted.

REFERENCES

Island references: the principal sources for all the island accounts have been my own observations but for the sake of conciseness they are not repeated under each island reference.

INTRODUCTION

J. M. Wilmshurst *et al.*, 2011, 'High precision radio carbon dating shows recent and rapid initial human colonisation of East Polynesia', *Proceedings of the National Sciences of USA 100 Academy* 5: 1815–20.

A. L. Pedersen, 2009, 'Formation of a bird community on a new island, Surtsey Iceland', *Surtsey Research* 12: 133–48.

ASCENSION ISLAND

Olivia Renshaw pers. com.

W. R. P. Bourne, N. P. Ashmole, and K. E. L. Simmons, 2003, 'A new subfossil night heron and a new genus for the extinct rail from Ascension Island', *Ardea* 91(1): 45–51.

Ascension Island Conservation Department news reports.

B. J. Godley, A. C. Broderick, and G. C. Hays, 2001, 'Nesting of green turtles (Chelonia mydas) at Ascension Island, South Atlantic', *Biological Conservation* 97(2): 151–8.

R. Huxley, 1999, 'Historical overview of marine turtle exploitation, Ascension Island', *Marine Turtle Newsletter* 84: 7–9.

J. Ripple, 1996. *Sea Turtles*, West Vancouver: Voyager Press Inc. World Wide Library.

CHINIJO ARCHIPELAGO

J. H. Flint and G. L. Morgan, pers. com.

D. A. Bannerman, 1922, *The Canary Islands, their history, natural history and scenery*, London: Gurney and Jackson.

R. R. Lovegrove, 1971, 'BOU supported expedition to Northeast Canary Islands', *Ibis* 113(2): 269–72.

Fernando de Noronha

H. N. Ridley, 1890, *Natural History of Fernando de Noronha*, London: Linnean Society.
Armando Santos, TAMAR, pers. com.
Natalia Menelan, Projeto Golfinho Rotador, pers.com.
Marine Turtle Newsletter 2007, 116:26.

Guam

Haldre Rogers, University of Washington, pers. com.
Isaac Chellman and Kaitlin Mattos, University of Washington, pers. com.
Daniel Vice, US Dept of Agriculture, pers. com.
Suzanne Medina, University of Guam, pers. com.

Halfmoon Island

R. Roberts, 2003, *Shipwrecked at the Top of the World*, New York: Time Warner, 169–204.
Palle Uhd Jepsen, pers. com.

Ile aux Aigrettes

Vikash Tatayah, pers. com.
Mauritian Wildlife Foundation factsheets *Red List of Threatened Species*.
Birdlife International, 2008, *Falco punctatus* in IUCN Red List of threatened species.
Birdlife species factsheets.

Jan Mayen

Susan Barr, pers. com.
S. Barr, 1987, *Norway's Polar Territories*, Oslo: Aschehoug.

Mingulay

B. Buxton, 1995, *Mingulay*, Edinburgh: Birlinn Press (principal reference).
H. Haswell-Smith, 2008 edition, *The Scottish Islands*, Edinburgh: Canongate.

Mykines

K. Williamson, 1948, *The Atlantic Islands*, London: Collins.

Brathay Exploration Group reports.
Katarina Johannessen, in litt.

PICO

B. Venables, 1968, *Baleia!* London: Bodley Head.

SAN BLAS ISLANDS

Personal observations only.

ST KILDA

A. Fleming, 2005, *St Kilda and the Wider World*, Oxford: Windgather Press.
T. Steel, 1975, *The Life and Death of St Kilda*, London: Harper Collins.

See also:
J. Harden and O. Lelong, 2011, *Winds of Change. The Living Landscape of Hirta, St Kilda*, Edinburgh: Society of Antiquaries of Scotland.
D. Gilles and J. Randall, 2010, *The Truth about St Kilda*, Edinburgh: John Donald.
C. Maclean, 1972, *St Kilda—Island at the Edge of the World*, London: Tom Stacey Ltd.
D. Quine, 1988, *St Kilda Portraits*, published by the author.

ST PETER AND ST PAUL ROCKS

Charles Darwin, 1860, *Journal of Researches into the Natural History and Geology of the countries visited during the voyage of H.M.S. Beagle round the world*, London: Murray, 8–9.
H. G. Smith *et al.*, 1974, 'A biological survey of St Paul's Rocks in the equatorial Atlantic Ocean', *Biological Journal of the Linnean Society* 6: 89–96.

GREAT SKELLIG

D. Lavelle, 1976, *Skellig*, Dublin: O'Brien Press.

SOLOVETSKI ISLANDS

R. Robson, 2004, *Solovki*, New Haven, CT: Yale University Press.

SOUTH GEORGIA

R. Burton, 2005, *South Georgia*, Government of South Georgia and pers. com.
I. B. Hart, 2001, *Pesca*, Salcombe, Devon: Aidan Ellis.
A. Martin, (S. Georgia Habitat Restoration) in litt.

See also

K. Crosbie and S. Poncet, 2005, *A Visitors Guide to South Georgia*, Princeton, NJ: Princeton University Press.

TRISTAN DA CUNHA

M. Holdgate, 1958, *Mountains in the Sea*, London: Travel Book Club.

J. H. Flint, pers. com.

J. H. Flint, 2011, *Mid Atlantic Village*, private publication.

P. Ryan, 2008, 'Important Bird Areas: Tristan da Cunha and Gough Island', *British Birds* 101(11): 586–606.

L. B. Holthuis, 1963, *Marine Lobsters of the World*, New York UN Food and Agriculture Organisation, Species catalogue vol. 3.

TUAMOTU ARCHIPELAGO

G. L. Morgan and I. T. Williams, pers. com.

Gabrielle Coulombe, pers. com.

VIGUR

Salvar Baldursson, in litt.

WRANGEL ISLAND

W. L. McKinlay, 1978, *Karluk*, Panther Books (Granada).

A. Stenensen, 1966, *North*, 20–7.

P. G. Kevan and J. D. Shorthouse, 2003, 'Behavioural thermoregulation in High Arctic butterflies', paper 44: Studies in Arctic Insects, Entomological Research Institute, University of Alberta.

IUCN World Heritage citation.

R. K. Headland, in litt.

INDEX

Bold entries refer to illustrations

219